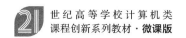

21 世纪高等学校计算机类
课程创新系列教材·微课版

Android
项目开发基础与实战

微课视频版

夏江 / 编著

清华大学出版社
北京

内容简介

本书结合作者多年讲授"Android 应用软件开发"课程的教学经验，融入 Android 开发领域新技术，较为全面地介绍了 Android 应用软件开发的相关知识点和开发技巧。全书共分 5 章，分别介绍了 Android 开发环境、Android 项目及 Java 基本概念、Android 常用布局、Android 常用控件和其他常用编程技术。本书相关案例以 Android Studio 为开发环境，尽可能使用最新版本 SDK 命令替代弃用命令，并对控件的版本变化做大致的介绍。本书全面考虑了本科教学的特点，结合作者开发的案例库教学辅助系统，通过精心设计的案例和详尽的讲解、演示，让读者感受体验式教学的魅力。

本书适合 Android Studio 开发人员、Android 的入门读者、高等学校学生使用，还可以作为高等院校、相关培训机构的教学用书。

本书封面贴有清华大学出版社防伪标签，无标签者不得销售。
版权所有，侵权必究。举报：010-62782989，beiqinquan@tup.tsinghua.edu.cn。

图书在版编目（CIP）数据

Android 项目开发基础与实战：微课视频版/夏江编著. —北京：清华大学出版社，2022.1
21 世纪高等学校计算机类课程创新系列教材：微课版
ISBN 978-7-302-59182-5

Ⅰ. ①A… Ⅱ. ①夏… Ⅲ. ①移动终端－应用程序－程序设计－高等学校－教材 Ⅳ. ①TN929.53

中国版本图书馆 CIP 数据核字（2021）第 187093 号

责任编辑：陈景辉　张爱华
封面设计：刘　键
责任校对：徐俊伟
责任印制：刘海龙

出版发行：清华大学出版社
　　　网　　址：http://www.tup.com.cn，http://www.wqbook.com
　　　地　　址：北京清华大学学研大厦 A 座　　邮　　编：100084
　　　社 总 机：010-62770175　　　　　　　　邮　　购：010-83470235
　　　投稿与读者服务：010-62776969，c-service@tup.tsinghua.edu.cn
　　　质量反馈：010-62772015，zhiliang@tup.tsinghua.edu.cn
　　　课件下载：http://www.tup.com.cn，010-83470236
印 装 者：三河市天利华印刷装订有限公司
经　　销：全国新华书店
开　　本：185mm×260mm　　印　张：18　　字　数：448 千字
版　　次：2022 年 1 月第 1 版　　　　　　　　印　次：2022 年 1 月第 1 次印刷
印　　数：1～1500
定　　价：59.90 元

产品编号：093316-01

前　言

　　Android作为目前智能设备的主流操作系统已覆盖平板电脑、手机、车载设备和智能电视等硬件设备，相关App应用也层出不穷。本书以最新版本的Android Studio为开发环境，详细讲解Android Studio的使用及Android应用程序开发技巧。

本书主要内容

　　本书以Android Studio开发环境配上多语言程序设计案例库教学辅助系统，以翔实的案例对Android应用开发的相关知识点进行循序渐进的讲解。

　　本书共分5章，内容编排如下。

　　第1章为熟悉Android开发环境，主要介绍Android开发环境搭建及Android Studio和案例库教学辅助系统的使用。

　　第2章为熟悉Android项目及Java基本概念，以向导建立的项目为基础，介绍Android项目的结构、相关代码文件和配置文件的作用及相互关系。通过案例对开发Android应用密切相关的Java概念做了较为深入的讲解。

　　第3章为Android常用布局，通过相关案例的介绍，读者可了解不同布局的特点，学会综合各种布局完成复杂的界面设计。本章对约束布局进行了详细的介绍。

　　第4章为Android常用控件，以案例为出发点，主要介绍Android开发设计中常用控件的属性、方法，对不同时期Android版本的相关控件外观和特性进行了较为全面的讲解。

　　第5章为其他常用编程技术，介绍了Intent、Activity、Menu、Service和数据库等内容，让开发人员对Android的综合应用开发有更深入的了解。

本书特色

　　（1）本书以本科教学为出发点，以长期教学中总结、归纳的精选案例为基础，结合教学辅助系统软件，打造了一个结合教学、演示和学习的案例库综合教学平台。

　　（2）将以讲解命令为主的方式转变为以阅读、理解代码为主的方式，让读者能从不同的角度认识Android的开发与应用。

　　（3）以基础知识点与案例相结合的方式，由浅入深、循序渐进地讲解知识点，代码注释详尽，便于读者将案例复制到Android Studio开发环境运行，同时提供在纯文本编辑器中对案例源码进行查询和修改。

　　（4）本书内容的组织、讲解及案例安排契合本科生的教学特点，注重内容质量，致力于服务教学和实现学生技能培训的目标。案例库支持使用者自行扩建，也为学生后续课程设计和毕业设计提供强有力的支持。

　　（5）由于Android版本更新较快，SDK的变化也较大，本书对相关控件演示最新变化时，也会讲解相关控件的特点、各版本差异、变迁和开发注意事项，让读者对Android开发有更全面的了解。

配套资源

为便于教学,本书配有230分钟微课视频、源代码、教学辅助软件、教学大纲、教学进度表、实验指导书。

(1) 获取教学视频方式:读者可以先扫描本书封底的文泉云盘防盗码,再扫描书中相应的视频二维码,观看教学视频。

(2) 获取案例库的源代码和教学辅助软件方式:先扫描本书封底的文泉云盘防盗码,再扫描下方二维码,即可获取。

源代码

教学辅助软件

(3) 其他配套资源可以扫描本书封底的课件二维码下载。

读者对象

本书适合Android Studio开发人员、Android的入门读者、高等学校学生使用,还可以作为高等院校、相关培训机构的教学用书。

本书的编写参考了诸多相关资料,在此对这些资料的作者表示衷心的感谢。

限于作者水平和时间仓促,书中难免存在疏漏之处,欢迎读者批评指正。

作　者

2021年11月

目 录

第 1 章 熟悉 Android 开发环境 ·· 1

1.1 Android 开发环境搭建 ·· 1
    1.1.1 Android IDE 选择 ·· 1
    1.1.2 Android Studio 的安装与配置 ··· 1
1.2 建立并运行 HelloAndroid ·· 9
1.3 Android Studio 项目结构 ·· 13
    1.3.1 app 目录 ·· 14
    1.3.2 Gradle Scripts ·· 15
1.4 Android Studio 常用功能 ·· 16
    1.4.1 Settings ··· 16
    1.4.2 Manage IDE Settings ·· 17
    1.4.3 Project Structure ··· 17
    1.4.4 Sync Project with Gradle Files ·· 18
    1.4.5 Bookmarks ·· 18
    1.4.6 Override Methods ··· 19
    1.4.7 Comment ··· 20
    1.4.8 Reformat Code ·· 21
    1.4.9 Build APK ··· 21
    1.4.10 Rename ·· 22
    1.4.11 Rename File ··· 22
    1.4.12 运行、调试 ··· 23
    1.4.13 向模拟器传送文件 ··· 23
    1.4.14 常用快捷键 ··· 24
    1.4.15 总结 ··· 25
1.5 多种语言程序设计案例库教学辅助系统 ·· 25
    1.5.1 使用对象及环境 ·· 26
    1.5.2 术语和缩写词 ·· 26
    1.5.3 软件安装 ·· 26
    1.5.4 软件使用说明 ·· 26

第2章 熟悉 Android 项目及 Java 基本概念 ·········· 32

2.1 熟悉 Android 项目 ·········· 32
2.1.1 Layout ·········· 32
2.1.2 Java 文件 ·········· 35
2.1.3 AndroidManifest.xml ·········· 38
2.2 调试输出 ·········· 40
2.3 类和对象 ·········· 42
2.4 extends ·········· 45
2.5 implements ·········· 48
2.6 super ·········· 51
2.7 equals 与恒等号（==） ·········· 53
2.8 方法重载 ·········· 58
2.9 代码块 ·········· 60

第3章 Android 常用布局 ·········· 62

3.1 Android 长度单位 ·········· 62
3.2 线性布局 ·········· 65
3.3 边线和角 ·········· 68
3.4 layout_weight ·········· 70
3.5 绝对布局 ·········· 71
3.6 相对布局 ·········· 74
3.7 帧布局 ·········· 76
3.8 表格布局 ·········· 78
3.9 网格布局 ·········· 83
3.10 约束布局 ·········· 86
3.10.1 约束布局基础 ·········· 86
3.10.2 Barrier ·········· 92
3.10.3 Guideline ·········· 95
3.10.4 Group ·········· 97
3.10.5 Circle ·········· 98
3.10.6 Chain ·········· 99
3.11 Space 和 layout_margin ·········· 103

第4章 Android 常用控件 ·········· 105

4.1 TextView ·········· 105
4.1.1 TextView 的常用属性和方法 ·········· 105

- 4.1.2 theme 和 style ……………………………………………………… 109
- 4.1.3 layout_gravity 与 gravity ……………………………………………… 112
- 4.1.4 findViewById()与 viewBinding ………………………………………… 113
- 4.2 Button …………………………………………………………………………… 115
  - 4.2.1 单击监听器 …………………………………………………………… 116
  - 4.2.2 监听器复用 …………………………………………………………… 119
  - 4.2.3 长按单击监听器 ……………………………………………………… 119
  - 4.2.4 动态添加按钮 ………………………………………………………… 122
  - 4.2.5 自定义 DoubleClickListener 监听器 ………………………………… 124
- 4.3 EditText ………………………………………………………………………… 126
  - 4.3.1 设置和获取文本 ……………………………………………………… 126
  - 4.3.2 按键监听器 …………………………………………………………… 129
  - 4.3.3 触摸监听器 …………………………………………………………… 130
  - 4.3.4 焦点改变监听器 ……………………………………………………… 131
  - 4.3.5 文本选择 ……………………………………………………………… 135
  - 4.3.6 禁止弹出软键盘 ……………………………………………………… 136
  - 4.3.7 inputType 和 imeOptions …………………………………………… 138
- 4.4 Toast …………………………………………………………………………… 139
  - 4.4.1 显示文本 ……………………………………………………………… 139
  - 4.4.2 显示图片 ……………………………………………………………… 140
  - 4.4.3 显示图片和文字 ……………………………………………………… 141
- 4.5 RadioButton …………………………………………………………………… 142
  - 4.5.1 获取单选按钮选中项 ………………………………………………… 143
  - 4.5.2 清空单选按钮 ………………………………………………………… 146
- 4.6 CheckBox ……………………………………………………………………… 148
  - 4.6.1 基本功能 ……………………………………………………………… 148
  - 4.6.2 监听器 ………………………………………………………………… 150
  - 4.6.3 代码复用 ……………………………………………………………… 152
- 4.7 CheckedTextView ……………………………………………………………… 153
- 4.8 ImageView ……………………………………………………………………… 157
- 4.9 DatePicker ……………………………………………………………………… 160
- 4.10 DatePickerDialog ……………………………………………………………… 162
- 4.11 TimePickerDialog ……………………………………………………………… 167
- 4.12 CalendarView ………………………………………………………………… 168
- 4.13 SeekBar ………………………………………………………………………… 171
- 4.14 RatingBar ……………………………………………………………………… 174
- 4.15 NumberPicker ………………………………………………………………… 178

  4.15.1 NumberPicker 基本功能 ………………………………………………… 178

  4.15.2 显示文字的 NumberPicker ………………………………………………… 180

 4.16 ProgressBar ………………………………………………………………………… 182

 4.17 Spinner ……………………………………………………………………………… 187

 4.18 ToggleButton ……………………………………………………………………… 191

 4.19 Switch ……………………………………………………………………………… 195

 4.20 AutoCompleteTextView ……………………………………………………………… 196

 4.21 ScrollView 和 HorizontalScrollView ……………………………………………… 199

 4.22 TextClock …………………………………………………………………………… 201

 4.23 Chronometer ………………………………………………………………………… 202

 4.24 AlertDialog ………………………………………………………………………… 206

  4.24.1 带默认按钮的 AlertDialog ………………………………………………… 206

  4.24.2 列表的 AlertDialog ………………………………………………………… 208

  4.24.3 单选的 AlertDialog ………………………………………………………… 209

  4.24.4 复选的 AlertDialog ………………………………………………………… 211

  4.24.5 自定义控件 ………………………………………………………………… 213

第 5 章 其他常用编程技术 ……………………………………………………………… 216

 5.1 Intent ………………………………………………………………………………… 216

  5.1.1 Intent 的显式调用和隐式调用 …………………………………………… 216

  5.1.2 Intent 传值和取值 ………………………………………………………… 219

 5.2 Activity ……………………………………………………………………………… 223

  5.2.1 系统状态栏、标题栏和导航栏 …………………………………………… 223

  5.2.2 关闭 Activity ……………………………………………………………… 226

  5.2.3 生命周期 …………………………………………………………………… 227

 5.3 电话及动态授权 …………………………………………………………………… 231

 5.4 发送短信 …………………………………………………………………………… 234

 5.5 Menu ………………………………………………………………………………… 236

  5.5.1 构建菜单 …………………………………………………………………… 236

  5.5.2 响应菜单项单击 …………………………………………………………… 237

  5.5.3 ContextMenu ……………………………………………………………… 239

 5.6 Notification ………………………………………………………………………… 241

 5.7 Service ……………………………………………………………………………… 244

 5.8 Broadcast …………………………………………………………………………… 249

  5.8.1 静态注册 …………………………………………………………………… 249

  5.8.2 动态注册 …………………………………………………………………… 251

  5.8.3 多接收器接收普通广播 …………………………………………………… 252

	5.8.4	有序广播	253
5.9		SQLiteDatabase	255
5.10		SQLiteOpenHelper	259
5.11		数据库调试	261
5.12		SharedPreferences	265
5.13		精度问题	266
5.14		横竖屏	271
5.15		获取 App 信息	272

附录 A	综合实验	274
参考文献		275

第 1 章　熟悉 Android 开发环境

1.1　Android 开发环境搭建

1.1.1　Android IDE 选择

Android 是一种基于 Linux 内核的操作系统。在 Android 应用开发中需要搞清楚两个概念：JDK 和 SDK。JDK(Java Development Kit)是 Java 程序开发包。JDK 的版本号随着 Java 的版本变化而变更。由于版权问题，又有了 OpenJDK 分支。Android 开发支持 JDK 或 OpenJDK，考虑到与 Oracle 公司的侵权诉讼问题，Google 公司在新推出的 Android Studio 开发平台中内嵌了 OpenJDK。SDK(Soft Development Kit)在 Android 开发中特指 Android SDK，它提供了 Android API 库、开发工具构建、测试和调试应用程序，其版本号与 Android 平台版本相对应。对于 Android 开发者而言，SDK 版本号对应 API Level，如 Android 11.0(R)对应 API Level 30(简称为 API 30)。在 Android 的项目开发中有一个 build.gradle 配置文件，其中有 minSdkVersion(支持 App 运行的最低版本)、targetSdkVersion (支持 App 运行的主要适配版本)和 compileSdkVersion(编译 App 的版本，建议尽量用最新的版本，可以在编写代码时避免使用被弃用的命令)。以上三者的关系为：

$$minSdkVersion <= targetSdkVersion <= compileSdkVersion$$

目前，Android 开发的 IDE 工具主要有 Eclipse 和 Android Studio。在早期的 Android 开发中，Eclipse 是第一选择，但安装困难、运行速度慢等问题也困扰了很多开发人员。Android Studio 作为官方主推的 IDE 工具在开发者群体中的占有率逐年增加。Android Studio 是基于 IntelliJ IDEA 的定制版，Google 公司针对 Android 的开发做了一系列的优化和集成，其安装和使用也越来越便捷。Android Studio 的设置和操作习惯与 IntelliJ IDEA 相同，IDE 的配置文件甚至可以相互导入导出。华为鸿蒙系统(HarmonyOS)的 App 开发平台 HUAWEI DevEco Studio 也是基于 IntelliJ 平台定制的。Android Studio 拥有功能强大的 UI 编辑器、完善的插件管理、Gradle 构建工具、智能保存、智能补齐代码等功能，已成为 Android 应用软件开发人员的编程首选。

1.1.2　Android Studio 的安装与配置

视频讲解

Android Studio 自 2013 年 5 月份推出以来，其目标向着"更快、更有效率"方向不停升级。在 Android Studio 的安装和配置上也体现出简约化的特色。早期搭建开发环境要先单独安装 Java 环境，再进行 Android Studio 的安装，最后是 SDK 的下载和配置，这需要一个漫长的过程才能完成。现在只要从 Android Studio 网站(https://developer.android.google.cn/studio)下载安装包,按照安装向导即可完成 Android Studio 开发环境的搭建。安装完成后的 Android Studio 中已经内置 OpenJDK,用户不用再单独安装 JDK,也不用像很多资料所说的配置 JAVA_HOME 和 PATH 环境变量。最新版的 Android Studio 除了.android(放置 Android 的虚拟设备文件)和.gradle 目录在 Windows 的用户目录下,其他文件都放置在用户安装时指定的目录下。安装过程比较耗时的是下载相关的开发组件。Android Studio 基本上可以算是绿色软件,如果用户不想通过安装包安装 Android Studio,可以把已经安装好 Android Studio 的目录复制到目标计算机上,此时.android 和.gradle 目录可以不用复制,只要运行\bin\studio64.exe 并重建虚拟设备即可(使用 Android Studio 下载 Gradle 会比较慢,直接复制现成目录可节约时间)。

【注】 Android Studio 的 SDK 目录下有上万个文件,如果选择多个版本的 SDK,其文件数量将增加更多。为加快 Android Studio 的启动、调试和运行速度,建议 Android Studio 和 SDK 目录放置在 SSD(固态硬盘),同时将计算机的内存扩充至 16GB 以上。

第一次运行 Android Studio 时会显示欢迎界面。Welcome to Android Studio 界面如图 1-1 所示。

图 1-1　Welcome to Android Studio 界面

如果已经建立了 Android 项目并重启 Android Studio,则显示建立过项目的 Welcome to Android Studio 界面,如图 1-2 所示。其左侧显示最近建立的项目。

单击右下角的 Configure 下拉选项,弹出如图 1-3 所示的下拉菜单。

选择 SDK Manager 选项时弹出 SDK Platforms 选项界面,如图 1-4 所示,其中,在 Android SDK Location 文本框中设定 SDK 的目录。在 SDK Platforms 中选中相应 Android 版本,单击 OK 或 Apply 按钮将自动下载并安装选中的 Android SDK 平台相关软件。用户可选中多个版本的 Android SDK 以便在后续的项目编程调试中进行不同版本的

图 1-2 建立过项目的 Welcome to Android Studio 界面

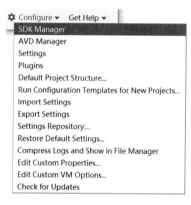

图 1-3 Configure 下拉选项菜单

运行测试。如果已经建立项目,也可以选择 File→Setting 菜单或 Tools→SDK Manager 菜单调出此配置界面。

如果选择 SDK Tools 选项卡,则显示 SDK Tools 选项界面,如图 1-5 所示。图 1-5 中所选项目分别为:

(1) Android SDK 生成工具。

(2) Android 虚拟设备,用于开发程序的运行、调试。如果不安装 Android Studio 自带虚拟设备也可以使用第三方虚拟设备或真实的 Android 设备。

(3) Android SDK 平台工具。

(4) Intel 虚拟设备加速器。对于使用 Intel CPU 的计算机,选中此项后将大幅提升虚拟设备的启动和运行速度。如果不使用 Android Studio 自带虚拟设备或非 Intel CPU 芯片的计算机可以不安装。使用 AMD CPU 芯片的计算机则应选中 Android Emulator Hypervisor Driver for AMD Processors(installer)复选框。

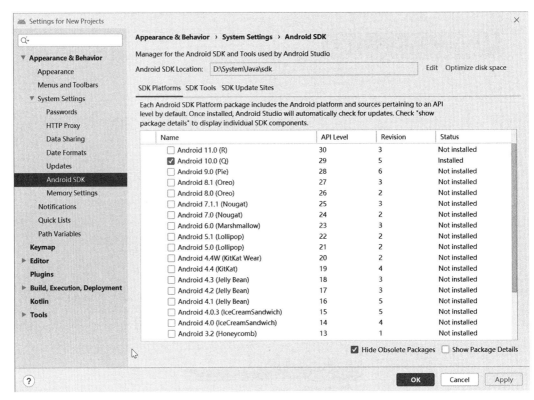

图 1-4　SDK Platforms 选项界面

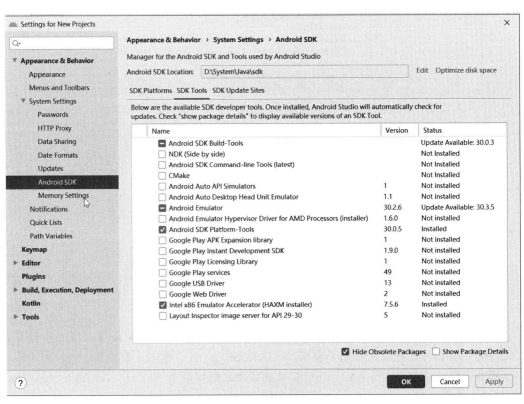

图 1-5　SDK Tools 选项界面

图 1-4 和图 1-5 中显示选中为 ☑ 代表已安装或准备下载安装,可在 Status 栏查看安装或升级状态,Version 栏显示安装包版本号。如果软件包前的复选框显示为 ➖,说明此选项有可升级的新版本,只要单击此复选框将其变更为选中 ☑,再单击 OK 按钮或 Apply 按钮,Android Studio 会将相应选项软件包升级到最新版本。

当选择图 1-3 中 AVD Manager 时会弹出虚拟设备管理器界面,如图 1-6 所示。在后续建立的项目中如果要在虚拟设备中调试、运行程序,就需要建立虚拟设备来模拟 Android 设备。

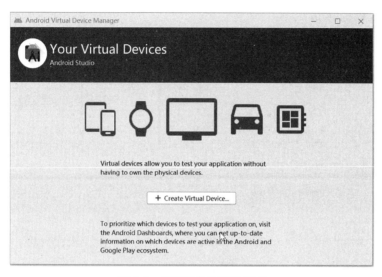

图 1-6 虚拟设备管理器界面

单击 Create Virtual Device 按钮转到选择虚拟设备界面,如图 1-7 所示,可以选择电视、手机、穿戴设备、平板和车载设备,默认选中手机。

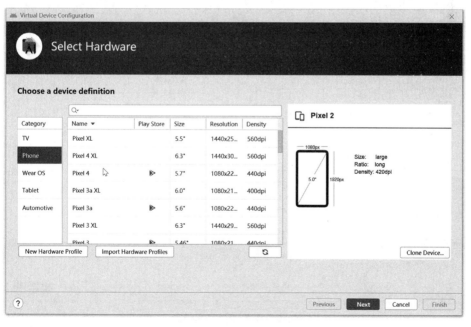

图 1-7 选择虚拟设备界面

选择虚拟设备或自定义一个虚拟设备，配置虚拟设备步骤如图1-8和图1-9所示。用户可自定义虚拟设备的名称、屏幕尺寸、分辨率、内存大小、横竖屏、前后摄像头以及传感器。

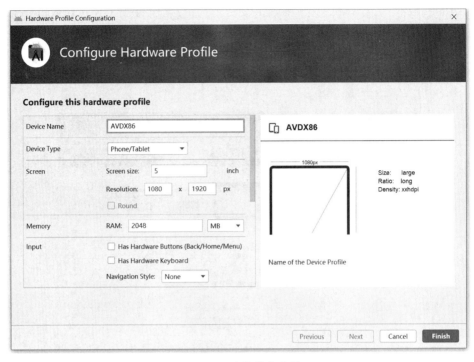

图1-8 配置虚拟设备步骤1

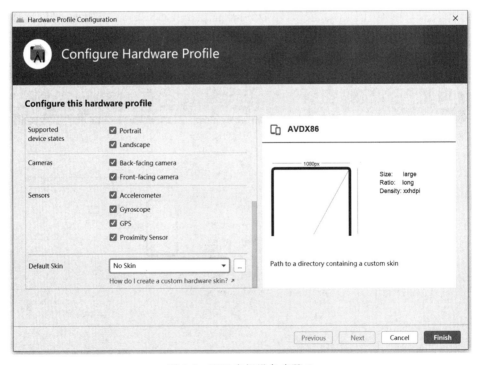

图1-9 配置虚拟设备步骤2

单击 Next 按钮转到选择虚拟系统界面,如图 1-10 所示。在其中选择要运行的 Android 系统版本。如果选择在 SDK Manager 中未下载的 Android 镜像版本,单击 Download 链接下载即可。

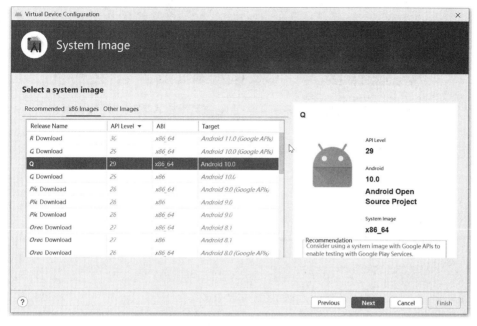

图 1-10　选择虚拟系统界面

单击 Next 按钮转到 Android 虚拟设备配置确认界面。如图 1-11 所示。单击 Show Advanced Settings 按钮可对虚拟设备的显卡、CPU 数量、内存、内置存储空间和 SD 容量进行设置。

图 1-11　Android 虚拟设备配置确认界面

单击 Finish 按钮完成虚拟设备配置,显示虚拟设备列表,如图 1-12 所示。单击列表右侧的下拉按钮▼,可对虚拟设备进行相关操作,例如复制、删除、冷启动虚拟设备等。Wipe Data 选项相当于恢复出厂设置,可以快速清空虚拟设备上的用户数据。

图 1-12　虚拟设备列表

单击▶按钮启动虚拟设备,显示启动单个模拟器,如图 1-13 所示。Tab 标签显示当前虚拟设备名称为 AVDX86(对应之前建立虚拟设备时取的名字)。如果建立了多个虚拟设备,此界面可以同时启动多个模拟器,如图 1-14 所示。

图 1-13　启动单个模拟器

图 1-14　启动多个模拟器

【注】 要同时运行多个模拟器必须先建立同样数量的虚拟设备(AVD),这与早期的一个虚拟设备可同时运行在多个模拟器上是不同的。如果是同时打开两个 Android Studio,则可以使用同一虚拟设备开启两个模拟器,只是这两个模拟器会同步显示,本质上还是同一个模拟器。

【故障现象】 有时在启动虚拟设备时会弹出如图 1-15 所示的故障。引发此故障有如下两种可能。

(1) 已经有一个同名的虚拟设备在模拟器上运行。

(2) 同名的虚拟设备所在模拟器上异常退出。

解决方法:

(1) 虚拟设备运行出错提示,如图 1-15 所示,将指定目录下的 *.lock 文件和子目录删除。

(2) 关闭或停止模拟器并重启 Android Studio。

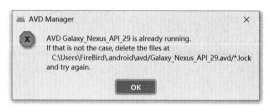

图 1-15　虚拟设备运行出错提示

新版本的 Android Studio 支持模拟器以窗口方式和停靠在 Android Studio 开发界面内部方式显示。模拟器显示界面的大小可由用户自行调节,与虚拟设备中的尺寸设置无关。

设置好 SDK 和 AVD 后就可以建立运行第一个 Android 项目程序了。

1.2　建立并运行 HelloAndroid

视频讲解

单击图 1-1 的 Create New Project 选项,进入选择项目模板,如图 1-16 所示。以常用的手机应用开发为例选择 Phone and Tablet 选项卡,选择 Empty Activity 创建一个简单 Android App 项目。

单击 Next 按钮进入项目配置,如图 1-17 所示。

将项目名称(Name)修改为 HelloAndroid,此名称即是 App 名称,建议以大写字母开头,用中文也是可以的,日常使用的支付宝、微信即为相应 App 的名称。包名称(Package name)一般使用域名的反向顺序格式,如 com.xiaj。当输入项目名称时,系统默认包名称为 com.example.helloandroid,即 com.example.＋项目名称。本书案例为方便查看代码,包名统一为 com.xiaj。在项目中将建立 com 目录下的 xiaj 子目录。包名在项目中涉及作用域的问题,这在后续案例中会进行详细讲解。开发项目所用语言选择 Java。虽然 Google 公司推荐首选 Kotlin,但 TIOBE 排名还在第 39 位(2021 年 2 月),上升比较缓慢(2017 年 6 月排第 43 位),相应的源码也少于 Java。Android Studio 建立的项目中可以同时包含 .java (Java 文件后扩展名)和 .kt(Kotlin 文件扩展名)文件。Kotlin 的优点是简化了 Java 中的编写代码方式,但同时也要看到 Java 也在不停地进行自我完善和升级。在 Minimum SDK 下

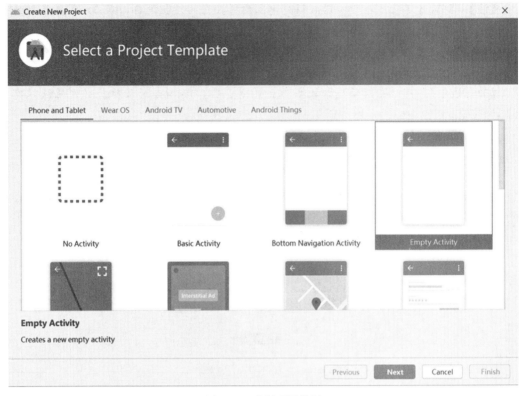

图 1-16　选择项目模板

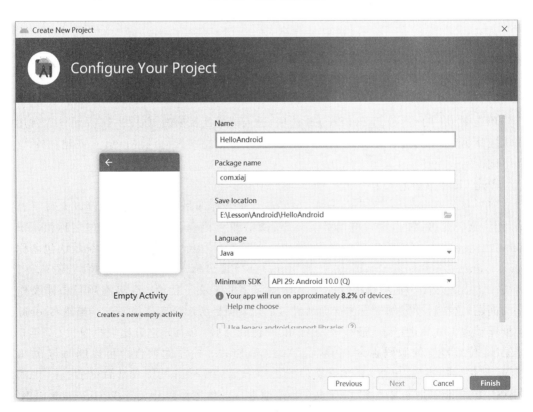

图 1-17　项目配置

拉列表框中选择项目要支持的最低版本 SDK。下拉列表框下方会显示兼容当前版本的设备数量百分比。单击 Help me choose 链接转到 API 版本分布，如图 1-18 所示。

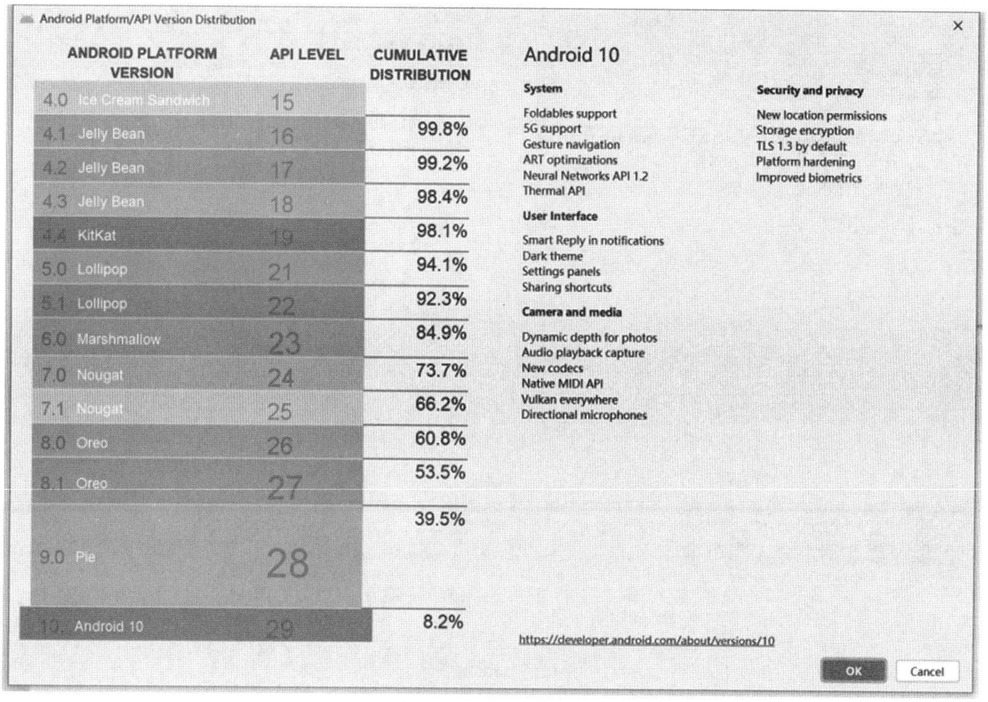

图 1-18　API 版本分布

【注】 项目保存路径中的各级目录名称必须为 ASCII，且不能出现"*"、空格，当然也就不能有中文了。新版本的 Android Studio 在图 1-17 中输入路径时会进行相应提示。用户在引入非自建的项目时需要特别注意这个问题。

单击 Finish 按钮，显示新创建的 HelloAndroid 项目，如图 1-19 所示。单击 ▶ 按钮，Android Studio 会自动启动模拟器，并运行默认的虚拟设备（后续运行将跳过这一步），将 HelloAndroid 程序编译后打包成 APK 安装文件传送到模拟器中，自动安装和运行程序，HelloAndroid 项目运行结果如图 1-20 所示。在整个项目生成和运行过程中未编写任何代码，Android Studio 已经构建好程序框架和基本代码。开发人员只要逐步添加功能代码就可以生成各式各样的 App。

如果想在真实的手机上运行 HelloAndroid 程序可按以下步骤处理。

（1）在手机的设置中找到"关于手机"选项，连续单击"版本号"选项栏 7 次（具体次数可能会因手机品牌不同而有差异）就可以激活"开发人员选项"（有的手机叫"开发者选项"）。

（2）进入"开发人员选项"，打开"USB 调试"。

（3）用 USB 线连接手机与计算机，在手机弹出的"USB 连接方式"界面中选择"传输文件"选项。

完成上述设置后，显示可连接的调试设备，如图 1-21 所示。Android Studio 的可连接设备下拉列表框中将显示已连接的手机和 AVD 虚拟设备。选中手机，单击 ▶ 按钮，当前项目自动安装到手机上并运行。

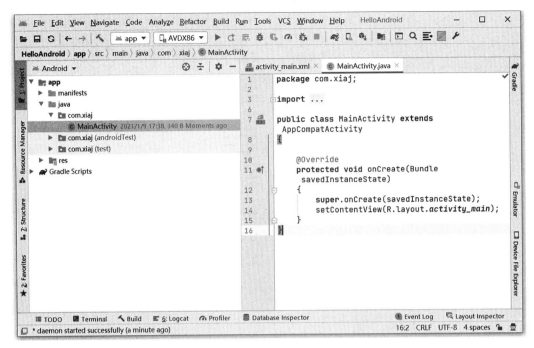

图 1-19　HelloAndroid 项目

图 1-20　HelloAndroid 项目运行结果

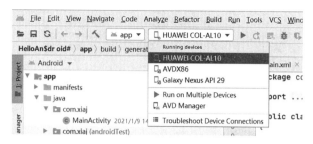

图 1-21 可连接的调试设备

如果在图 1-21 中选中 Run on Multiple Devices 选项,将弹出项目同时运行在多个设备的选项列表,如图 1-22 所示。选中相应设备,单击 Run 按钮,当前项目自动安装到所有选中的设备上并运行。此项功能为开发的软件在不同版本设备上进行兼容性调试带来极大的方便。

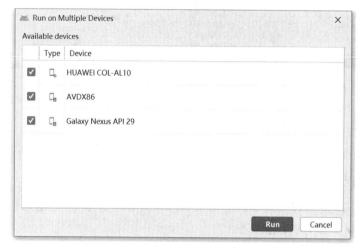

图 1-22 项目同时运行在多个设备的选项列表

1.3 Android Studio 项目结构

视频讲解

  Android Studio 项目结构视图如图 1-19 所显示。Android Studio 大致分为上部的菜单和快捷按钮区、左侧显示结构的导航功能区和右侧代码显示功能区。根据所显示的文档不同,右侧功能区可能会划分出更多的显示区域,如选中布局文件会显示源码视图、设计视图、组件树等。在左侧的导航功能区上部有一个下拉按钮,单击该按钮会显示不同的导航视图,默认选中 Android 导航视图,如图 1-23 所示。

  Android 导航视图显示常用布局文件、Java 文件及相关项目配置文件,如图 1-24 所示。各种导航视图显示的目录、文件会有差异,最接近实际项目保存目录结构的是 Project 导航视图。不论哪一种导航视图,都只是对实际保存项目目录和文件的选择性显示。

  Android Studio 的项目保存目录结构随着 Android Studio 版本的变化一直在不停地调整,但导航视图的结构变化不大。以下以最新版 Android Studio 的 Android 导航视图为例讲解 HelloAndroid 项目中包含的主要目录和文件。

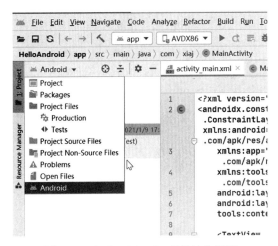
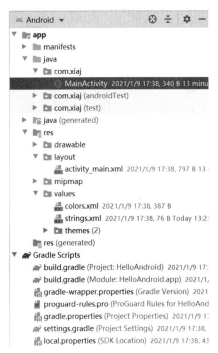

图 1-23 Android Studio 项目结构视图　　图 1-24 Android 导航视图

1.3.1　app 目录

布局文件、Java 源码以及编译以后的文档都放置在此目录下,它也是编程中经常使用的目录。app 目录下常见子目录如下。

1. manifests 目录

在 Android 项目中其实此目录是不存在的,实际指向的是 HelloAndroid\app\src\main 目录下的 AndroidManifest.xml 文件。文件中定义了项目的名称、主题、Activity 注册、Service 注册、Receiver 注册、权限注册等。在早期版本中甚至 SDK 的版本控制也放在此文件中(现在迁移到 HelloAndroid\app\build.gradle 文件中)。

2. java 目录

java 目录对应 HelloAndroid\app\src\main\java 目录。在 Android 导航视图的 java 目录中包含 com.xiaj 子目录。com.xiaj 是建立项目时的 package 名称,对应 com 目录和其下的 xiaj 子目录,实际对应的路径为 HelloAndroid\app\src\main\java\com\xiaj。使用 package 的目的是便于设定作用域。在 HelloAndroid\app\src\main\java\com\xiaj 目录下保存的是 Java 文件。

3. res 目录

此目录下保存的文件都属于资源文件。

(1) drawable:用于保存图形相关文件,对应项目的 HelloAndroid\app\src\main\res\drawable 目录和 HelloAndroid\app\src\main\res\drawable-v24 目录。早期版本还包含图形文件。

（2）layout：用于在 Android 设备上显示图形界面，在 Android 中定义为视图（View）。布局文件遵循 XML 定义规范，文件扩展名为.xml。在本例中，布局文件路径为 HelloAndroid\app\src\main\res\layout。

（3）mipmap：用于保存图形文件，文件扩展名为.png，实际对应 mipmap 开头的 6 个子目录，如 HelloAndroid\app\src\main\res\mipmap-hdpi、HelloAndroid\app\src\main\res\mipmap-anydpi-v26 等。不同目录下保存同名的不同分辨率的图片，以便不同分辨率的 Android 设备有相似的显示效果。

（4）values 目录：用于保存主题、颜色配置、字符串资源文件。在新版本的 Android Studio 中对字符串资源文件的定义中取消了<? version="1.0" encoding="UTF-8"? >这一行说明，如本例中 strings.xml 文件格式如下：

```
<resources>
    <string name="app_name">HelloAndroid</string>
</resources>
```

其含义为定义了一个名称为 app_name 的变量，其值为 HelloAndroid。定义的变量可在布局文件或 Java 文件中引用。如果变量 app_name 在多个文件中使用，并需要将其值改为"你好安卓"时，只需修改 strings.xml 中 app_name 的值就可，这将极大地方便后续的被引用值修改或者发布多语言版本。

1.3.2　Gradle Scripts

（1）build.gradle：用于 Gradle 构建管理的配置文件，一般采用生成向导默认生成内容即可。这里有两个 build.gradle 文件：Android 导航视图中为了区别，前者显示为"build.gradle(Project:HelloAndroid)"，保存的路径为 HelloAndroid\build.gradle。后者显示为 "build.gradle(Module:HelloAndroid.app)"，保存的路径为 HelloAndroid\app\build.gradle，其中包含 SDK 版本设置（compileSdkVersion、minSdkVersion、targetSdkVersion）、项目版本编号（App 版本升级依据）、项目版本名称以及项目使用到的相关本地依赖、库依赖和远程依赖（如 SQLiteStudio 直接查看项目中 SQLite 数据库的设置）。

（2）gradle.properties：Gradle 的全局配置相关文件。其中涉及的配置为：

org.gradle.jvmargs=-Xmx2048m -Dfile.encoding=UTF-8

它用于分配 JVM 堆内存大小和指定文本编码方式。

上述结构目录和文件只占实际项目的一部分，HelloAndroid 案例中实际有 174 个文件夹、458 个文件，相比早期版本的 438 个文件夹、860 个文件已经大幅缩减。对比新旧 Android Studio 版本的项目，主要变化是在编译所需文件和保存编译结果的 build 目录。如果再考虑 Android Studio 开发环境本身的 32 488 个文件（含 Windows 用户目录下的.android 目录和.gradle 目录），要想流畅地运行 Android Studio 项目，最好将 Android Studio 和开发项目放置到 SSD 中。

1.4 Android Studio 常用功能

Android Studio 的功能非常丰富,用户可以根据自己的使用习惯来修改或定制相应功能。因篇幅限制,本书只针对日常教学和编程中经常用到的功能进行讲解。

1.4.1 Settings

视频讲解

选择 File→Setting 菜单打开 Settings 界面,如图 1-25 所示。

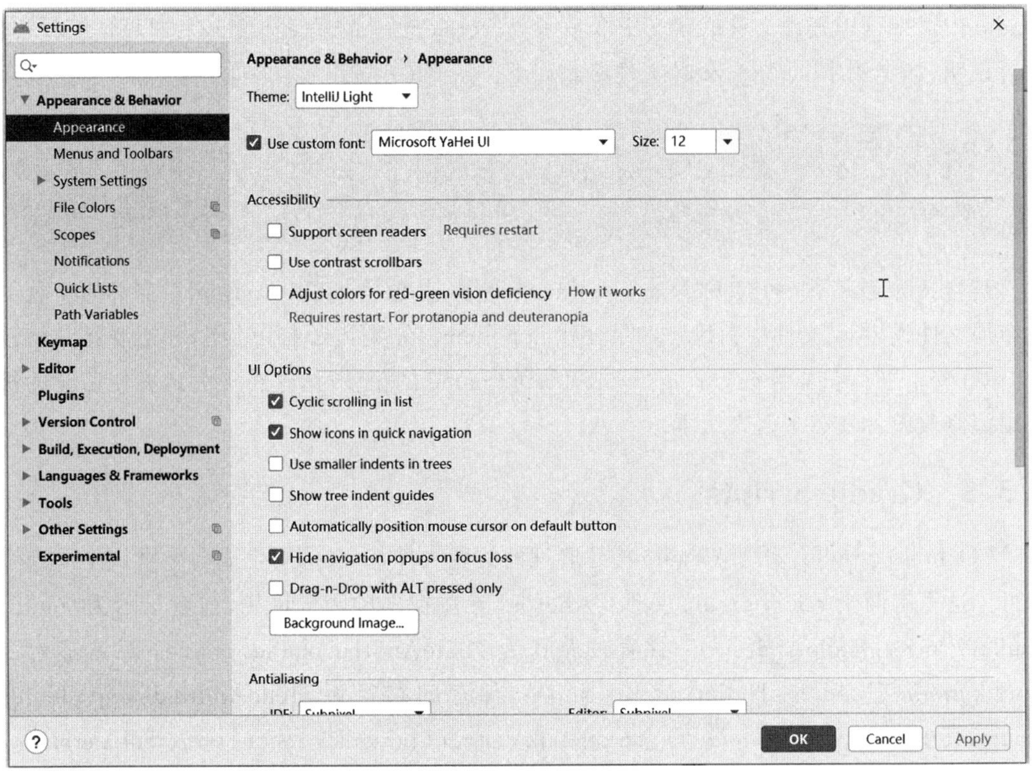

图 1-25 Settings 界面

可以根据个人喜好设置 Android Studio 开发环境主题、代码显示风格、智能提示、快捷键、插件、版本控制等。在使用 Android Studio 的过程中如果感觉某项功能不太便,可在此界面中修改相应配置解决自己的困扰。通过对 Settings 的设置和慢慢磨合,将 Android Studio 打造成真正适合于自己的 IDE 开发环境。

本书默认采用的 Keymap 风格为"Windows copy Proper Redo",书中列出的快捷键也是以此为基础的默认设置。Keymap 风格设置如图 1-26 所示。读者可选择自己习惯的快捷键映射风格,也可修改其中的快捷键映射。熟练使用快捷键能提高编程效率、改善编程体验。

Editor 选项可修改编辑器界面字体类型、字体大小、编码风格等,让代码更便于阅读。

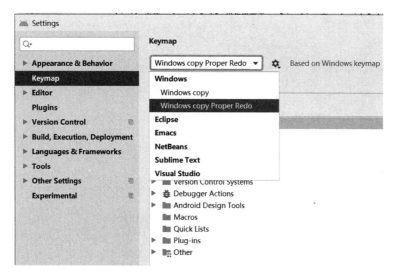

图 1-26　Keymap 风格设置

1.4.2　Manage IDE Settings

在 Settings 中设置好配置选项,如果想保留其配置可以选择 File→Manage IDE Settings→Export Settings 菜单,弹出导出配置界面,如图 1-27 所示。

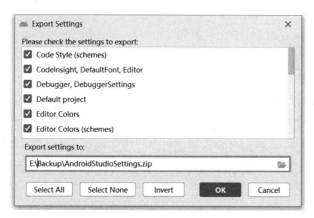

图 1-27　导出配置界面

选择需导出的配置选项,指定配置文件保存目录,单击 OK 按钮就可导出 Android Studio 相关配置。在其他 Android Studio、Intellij IDEA 或 HUAWEI DevEco Studio 开发软件中依次选择 File→Manage IDE Settings→Import Settings 菜单,可将之前导出的压缩配置文件导入相应的开发软件中,让上述几种开发软件的配置保持同步。

1.4.3　Project Structure

选择 File→Project Structure 菜单,弹出 Project Structure 界面,如图 1-28 所示。在 Android SDK location 中指定 SDK 存储路径。在 JDK location 中指定 JDK 路径,默认使用 Android Studio 内置的 JDK。

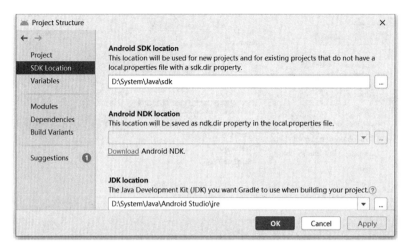

图 1-28 Project Structure 界面

1.4.4 Sync Project with Gradle Files

当 Android Studio 打开与当前版本不匹配的项目时,工具栏的 app ▼ (Run/Debug Configuration)会出现红色× app ▼ ,选择 File→Sync Project with Gradle Files 菜单会自动同步项目所需文件。如果还未解决问题,则重启 Android Studio 后大都能恢复正常。Android Studio 虽然一直在升级、改进,但还是会出现一些莫名其妙的故障,在查找不到故障原因时不妨重启 Android Studio 试试。

1.4.5 Bookmarks

选择 Navigate→Bookmarks→Toogle Bookmark 菜单(快捷键为 F11)时,对光标所在代码行添加书签标记 ✓。单击 Toggle Bookmark with Mnemonic 菜单项(快捷键为 Ctrl+F11)时,对光标所在代码行添加书签标记 ✓,同时弹出包含数字和字母的界面,选择相应数字或字母,书签标记 ✓ 变为带数字或字母的助记书签标记,两种 Bookmarks 如图 1-29 所示。

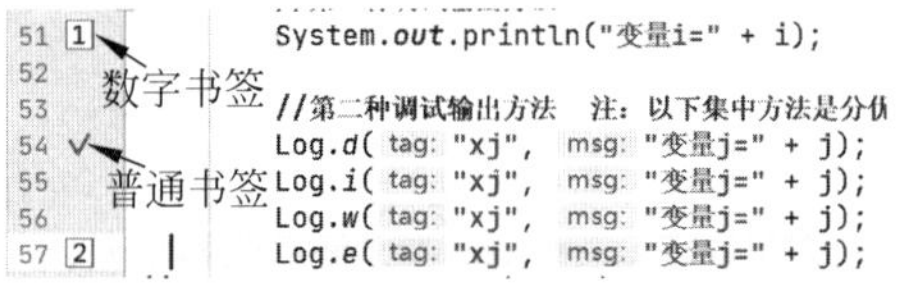

图 1-29 两种 Bookmarks

在已添加书签标记的代码行再次选择上述菜单或按相应的快捷键(F11/Ctrl+F11)都会取消书签标记。

当选择 Show Bookmarks 菜单(快捷键为 Shift+F11)时,弹出 Bookmarks 导航界面,如图 1-30 所示。开发人员可将编辑栏快速转到添加书签标记代码行。

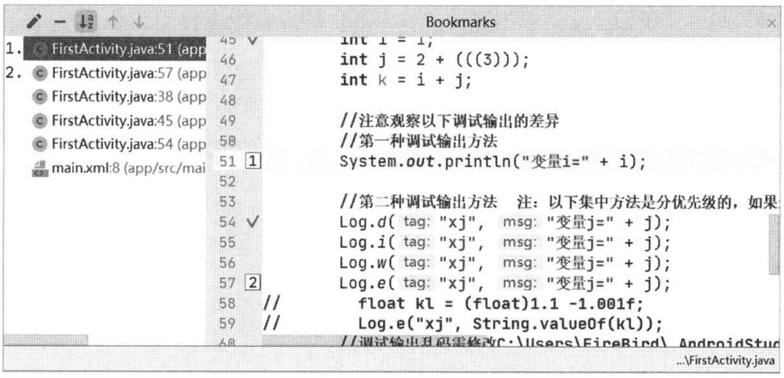

图 1-30　Bookmarks 导航界面

1.4.6　Override Methods

选择 Code→Override Methods 菜单(快捷键为 Ctrl＋O)，弹出 Override Methods 界面，如图 1-31 所示。

图 1-31　Override Methods 界面

当选择某一方法如 onRestart()，单击 OK 按钮后，Android Studio 自动在当前文档的 public class 类中添加 onRestart()方法的框架代码，开发人员只需在此框架下添加后续功能代码。自动添加的 onRestart()框架代码如下：

```
01    @Override
02    protected void onRestart()
03    {
04        super.onRestart();
05    }
```

为方便讲解，后续列出的代码前都会加上行号，实际代码中是没有的。

1.4.7 Comment

选择 Code→Comment with Line Comment 菜单（快捷键为 Ctrl+/）实现对当前行或选中的代码行加行注释或取消当前行已有行注释。

行注释是在选定的行前加"//"，例如：

```
//int i = 0;
```

行注释也可用于代码行之后，例如：

```
int i = 0;      //i：统计运行次数
```

选择 Code→Comment with Block Comment 菜单（快捷键为 Ctrl+Shift+/）实现对当前选中代码行加块注释或对已加块注释的代码块取消块注释。块注释是在选定的代码块前后加"/*"和"*/"，例如：

```
/* int sum = 0;
for (int i = 0; i < 10; i++)
{
    sum += i;
} */
```

除了以上两种注释以外，还有一种注释叫文档注释，以"/**"开始，以"*/"结束。文档注释可以被 JDK 提供的工具 javadoc 所解析，生成一套能够以网页形式体现程序的说明文档（选择 Tools→Generate JavaDoc 菜单可自动生成，目前版本需要文档编码转换为 GBK 才能正常生成网页，可使用命令行方式加入编码格式 UTF-8）。在 Android Studio 中使用文档注释的方法可以在调用此方法时显示文档注释内容。例如，已经编写好如下方法：

```
01    String setName(String name)
02    {
03        return "姓名：" + name;
04    }
```

在第 1 行上方输入"/**"然后按 Enter 键，Android Studio 自动生成如下文档注释：

```
/**
 *
 * @param name
 * @return
 */
```

修改文档注释如下：

```
/**
 * 本方法用于格式化用户姓名
 * @param name 用户姓名
 * @return 返回字符串"姓名："+ name
 */
```

需要注意，文档注释中包含@的行在添加字符时与原有内容之间要用空格间隔。当调用setName()方法时，将光标放置在setName上，按下快捷键Ctrl+Q（对应View→Quick Documentation菜单），弹出显示文档注释，如图1-32所示。智能提示中显示文档注释中的相关内容。开发人员在输入代码时可以很方便地查询代码文档。

图1-32 显示文档注释

【注】 常用注释标记及默认顺序：

@author（用于在类和接口的文档注释中指明作者）。

@version（用于在类和接口的文档注释中指明版本）。

@param（出现在方法或构造方法的文档注释中指明相关参数）。

@return（出现在有返回值的方法中）。

@deprecated（代表已弃用）。

1.4.8 Reformat Code

选择Code→Reformat Code菜单（快捷键为Ctrl+Alt+L），Android Studio将按照Settings中配置的代码风格自动对当前文档进行排版，包含空行数量、缩进、空格数量、花括号位置、折行方式等。

【注】 命令行内部合理的空格及命令行之间的合理空行（留白）可提高代码可阅读性。

1.4.9 Build APK

选择Build Bundle/APK→Build APK菜单，Android Studio会在项目的app\build\outputs\apk\debug目录下生成一个名为app-debug.apk的未签名APK文件，此文件可以提供给其他Android设备使用。选择Build→Select Build Variant菜单可以选择生成APK的debug版本或release版本，两者的区别在于前者包含调试信息，后者是进行了优化的发布版本。

【注1】 选择Run→Run'app'菜单也会在app\build\outputs\apk\debug目录下生成app-debug.apk文件，此时有一些文件未封装到APK中，因此这个安装包文件要比Build Bundle/APK→Build APK生成的安装包文件小，只能在调试的设备上安装运行。

【注2】 APK文件其实就是ZIP文件，感兴趣的读者可以将APK文件扩展名改为.ZIP，解压后查看其中的目录结构和文档内容（其中的文件基本都是重新编码的）。

1.4.10 Rename

选择 Refactor→Rename 菜单(快捷键为 Shift+F6)可以对选中的标识符(含类名、变量名、方法名等由用户命名的名字)进行修改,同一作用域内的同名标识符也将自动同步变更。不同作用域的同名标识符不受影响。当需变更的标识符也出现在字符串中时,会弹出 Rename 上下文菜单,如图 1-33 所示。Rename code occurrences 菜单将只修改同一作用域的同名标识符。Rename all occurrences 菜单将修改同一作用域的所有同名标识符(包括字符串中与标识符同名的字符串)。

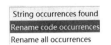

图 1-33　Rename 上下文菜单

本功能也可用于修改包名的最后一段,修改内容包括文件中引用的包名和对应的目录名称。在 Android Studio 左侧导航栏中选择 com.xiaj,选择 Rename,显示修改包名界面,如图 1-34 所示。

修改包名的最后一段,单击 Refactor 按钮,在 Android Studio 下方显示执行修改包名重构界面,如图 1-35 所示;界面上方显示需要修改的位置,单击 Do Refactor 按钮,自动修改包名重构,同时修改对应的目录名称。

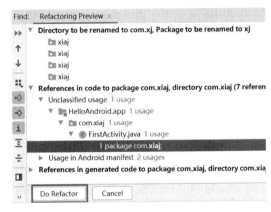

图 1-34　修改包名　　　　　图 1-35　执行修改包名重构界面

可能会出现导航栏的包名称还是未变,但 Java 文件中包名已被修改的情况,这是 Android Studio 中存在的一个缺陷。选择 File→Sync Project with Gradle Files 菜单或重启 Android Studio 后就能正常显示。

1.4.11 Rename File

选择 Refactor→Rename File 菜单可以对当前文件或导航栏视图中选中的文件进行重命名,弹出的 Rename File 界面如图 1-36 所示。由于重命名涉及对本文件引用的文档,需要在图 1-36 的 Scope 中指定对重命名文件进行引用的文件搜寻范围。

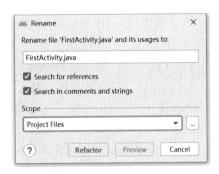

图 1-36　Rename File 界面

1.4.12 运行、调试

Run 菜单中有运行、调试等菜单，由于这些功能使用较为频繁，更便捷的方式是使用对应的工具栏按钮或快捷键。

(1) Run 按钮 ▶（快捷键 Shift+F10）：如本例选择 Run→Debug Configurations 菜单命名 app，则相应菜单或 Run 按钮的弹出提示为 Run'app'。当成功运行此项目以后，此按钮变为 ↻，同时 ◳（Apply Changes and Restart Activity）和 ⮂（Apply Code Changes）按钮也由灰色禁用状态变为绿色可用状态。注：本书为黑白印刷，具体颜色请参看相应操作界面，下同）

以上三个按钮的区别如下。

① 单击 ▶ 按钮会依次运行后续几个步骤：重新编译生成安装包、传送安装包到调试设备、重新安装程序、重新启动程序运行。单击此按钮，程序的运行速度是在单击这三个按钮中最慢的，但能真实反映程序的所有变化。

② 单击 ◳ 按钮会将变更代码传递到调试设备、重新启动程序至单击 ◳ 按钮时的界面。例如，在项目中建立两个 Activity，当启动时显示第一个 Activity，单击其中的按钮转到第二个 Activity，第二个 Activity 显示文本 OK，这时将程序代码 OK 修改为 YES，单击 ◳ 按钮，Android 设备上直接跳过第一个 Activity，在第二个 Activity 中显示修改后的文本 YES。此按钮可简化有多个操作步骤的程序调试，加快程序测试速度。单击此按钮，程序的运行速度在单击这三个按钮中是中等的，能反映绝大多数程序代码的变化。

③ 单击 ⮂ 按钮会将变更的代码传递到调试设备，但程序和相应界面（Activity）都不重启，后续程序的运行会按变更代码后的状态运行。部分影响 Activity 运行的代码改变会弹出 Apply Code Changes 按钮运行错误提示，如图 1-37 所示。此时可单击提示中链接或单击前两个按钮重新运行程序。单击此按钮，程序的运行速度是在单击三个按钮中最快的，但由于没有重启 Activity，不能完全正确反映实际程序运行变化。

三个按钮的实际运行差异可参见"5.2.3 Activity 生命周期"中的案例。

(2) Debug 按钮 🐞：设置断点。如图 1-38 所示，单击将执行中断的可执行代码行号旁边空白位置，出现红色断点图标，单击 🐞 按钮，程序将停在断点位置，当前代码行背景色反选。

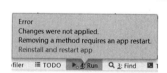

图 1-37 Apply Code Changes 按钮运行错误提示

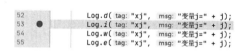

图 1-38 设置断点

Debug 界面如图 1-39 所示，方框中的按钮可实现各种步进调试以及查看变量和自定义表达式。没有不犯错的开发人员，在编程过程中多练习、多实践是排除程序故障的捷径。

1.4.13 向模拟器传送文件

选择屏幕右侧 Device File Explorer 窗口选项卡，弹出 Device File Explorer 界面，如图 1-40 所示。选中相应目录并右击，在弹出的快捷菜单中选择 Upload 菜单，实现计算机向模拟器上传文件。在弹出的快捷菜单中选择 Save As 菜单，实现模拟器向计算机下载文件。

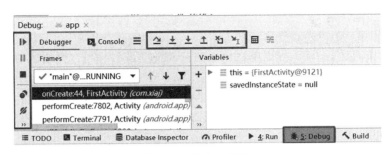

图 1-39　Debug 界面

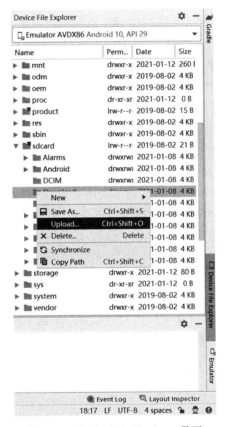

图 1-40　Device File Explorer 界面

1.4.14　常用快捷键

（1）Ctrl+E：弹出最近打开的文件列表。

（2）Ctrl+/：给当前行或选中的代码行添加或取消注释。

（3）Ctrl+Q：弹出光标所对应标识符的说明文档。

（4）Ctrl+Shift+向上（或下）箭头：将当前行或选中的代码块上移（或下移），移动范围限制在当前所在类或方法内部。

（5）Alt+Shift+向上（或下）箭头：将当前行或选中的代码块上移（或下移），移动范围不受限制。

(6) Ctrl+D：复制当前行或选中的代码块并粘贴在选定代码之后。

(7) Shift+Delete：删除当前行或选中的代码块。

(8) Shift+F6：同步重命名标识符或文件名。

(9) Ctrl+空格：匹配代码智能提示。因与输入法切换快捷键冲突，最好选择 Settings→Keymap 菜单修改快捷键。

(10) Ctrl+Shift+Backspace：返回上次编辑的位置。

(11) F11/Ctrl+F11：添加/删除书签标记或助记书签标记。

(12) Ctrl+Alt+L：将当前文件按 Settings 中格式重排代码(含 XML、Java 文件)。

(13) Alt+Enter：导入光标所在字符串对应的类，将自动生成一条 import 命令。

(14) Double Shift(按两次 Shift 键)：全局搜索。

(15) Ctrl+鼠标左键(Ctrl+B)：跳转到定义类、方法、成员变量的源码文件。

(16) Ctrl+Alt+O：删除没有被引用使用的包。

(17) Shift+F10/F9：Run/Debug。

以上快捷键会因所选 Keymap 主题的不同而有差异。选择 Android Studio 的 Help→Keymap Reference 菜单可获取默认的快捷键列表。开发人员也可以选择 File→Settings 菜单，在弹出的设置界面中选择 Keymap，寻找相应的动作事件修改快捷键。

1.4.15 总结

以上只是 Android Studio 众多功能中常用的一部分，还有很多设置可提高开发人员的编程效率。在菜单、工具栏和上下文菜单中都可提供有相同的功能。熟悉这些功能的最快捷方式就是多练习。对于个人常用的功能可以将其移到工具栏中。最后使用 File→Manage IDE Settings 菜单的 Import Settings 和 Export Settings 菜单项将 Android Studio 的设置参数进行导入和导出。

1.5 多种语言程序设计案例库教学辅助系统

视频讲解

在程序设计开发课程的传统讲授中教师常以 PPT 演示为主，对代码较多的案例往往需要多页 PPT 才能展示，且因字体大小、颜色、对齐等问题导致可读性降低。同一案例中代码变更对运行结果的变化也很难直观展示，更不用提实时运行、中断等演示问题。

学生在学习和复习过程中对相关案例查找和运行也需要重复多次复制、粘贴和修改源码及项目配置文件方能正常运行。

"多种语言程序设计案例库教学辅助系统"以方便教师的备课、教学和演示以及提高学生的复习效率为设计出发点，同时兼顾 Android、Java、Python、C 和 C++等多门程序设计开发课程的案例教学，突出简单、易用、实用和快捷的特色。

用户在使用本软件时能更方便地实现不同项目案例间的快速切换、演示。对开发环境升级导致的配置修改可实现一次修改就能覆盖所有项目案例的特点。案例库中源码也可直接推送到第三方文本编辑器中显示而无须安装 Android Studio 开发环境。

本软件经 Visual Studio 2019 版本 16.6.0 编译生成，支持环境 Microsoft .NET Framework 4.8.03752。相关教学辅助系统软件及案例库会随书提供给读者使用。

1.5.1 使用对象及环境

对于教师，本软件可用于 Android、Java、Python、C 和 C++ 等程序设计开发课程的教学案例建立、备课和实际编程环境下相关案例的运行、演示。

对于学生，本软件提供快速查看案例源码功能和实际编程环境下相关案例的快速演示功能。

软件可运行在安装 Windows 7/8/10、Windows Server 2003/2008/2012/2016/2019 等操作系统的个人计算机或服务器上。

1.5.2 术语和缩写词

(1) 案例库目录：此目录必须建立在 Lesson 目录或子目录下，用于保存相应语言程序设计课程的所有案例。每个案例都保存在单独的一个子目录下，相应子目录用中文或英文字符标识案例名称(字符只受 Windows 目录命名限制，与开发语言的目录字符限制无关)。

(2) 运行目录：相应开发语言的 IDE 软件运行案例的目录，包含案例项目的源码文件、项目管理文件、项目运行配置文件等。

(3) 快速演示：包含以下两个方面。

① 将选中的案例库中的项目复制到运行目录，开发环境可以直接运行新案例项目；

② 将选中案例库中的项目源码和配置文件直接按文本文件方式显示。

1.5.3 软件安装

本软件为绿色软件，用户将文件 XCopy.exe 复制到自定义的目录下就可以正常运行。用户使用软件的设置功能后，相关配置会自动保存在同一目录。

1.5.4 软件使用说明

(1) 运行主界面。

启动应用程序后直接进入主界面。如果是第一次使用，则主界面(未设置目录信息)如图 1-41 所示，相关内容显示为空。

如果已经通过"工具"→"选项"功能设置了相关运行目录，主界面(已设置目录信息)如图 1-42 所示。

(2) 如果未设置软件运行的相关目录，选择"工具"→"选项"菜单，如图 1-43 所示。

(3) 弹出"选项"对话框，如图 1-44 所示。

(4) 对话框中列出了常见的几种开发语言，后续会根据用户需求增加开发语言种类。每一种开发语言包含了案例库目录和运行目录。

(5) 单击相应目录文本框右侧的按钮，弹出选择相关目录的对话框，如图 1-45 所示。

(6) 为防止对其他文件夹的误操作，软件要求选择的目录必须在 Lesson 目录下。如果用户选择的目录路径中没有包含 Lesson 目录，会弹出路径选择错误提示，如图 1-46 所示。

(7) 在图 1-45 所示的对话框中单击"确定"按钮，选择的相应目录将显示在"选项"对话框的对应目录文本框中。如果同时需要多种开发语言的目录设置，重复刚才的目录设置即可。

图 1-41 主界面（未设置目录信息）

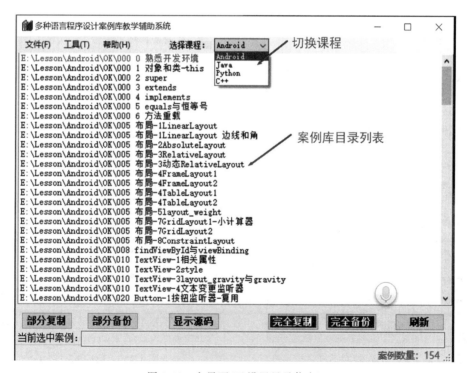

图 1-42 主界面（已设置目录信息）

熟悉 Android 开发环境

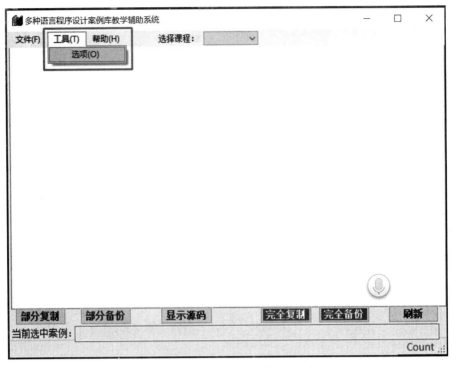

图 1-43　选择"工具"→"选项"菜单

图 1-44　"选项"对话框　　　　　　图 1-45　选择相关目录

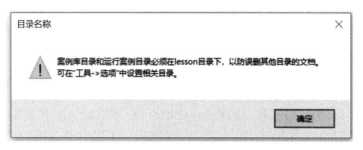

图1-46　路径选择错误提示

(8) 单击"选项"对话框中的"确定"按钮,软件将设定的目录路径自动保存在Config.ini文件中,并回到主界面。

(9) 在主界面的"选择课程"下拉列表框中可自由切换不同的开发语言课程。重启软件时会自动记忆上次退出时选中的开发语言课程。

(10) 按钮功能详解。

① 案例库目录显示区下方有6个按钮,其中4个用于复制和备份。对于Android应用开发,每个项目往往都有700个以上的文件和目录,其中主要是项目管理文档和编译运行过程文档,源码文档在其中占比不到1%。而不同案例的项目管理文档往往都是相同的。为减少案例库的空间占用,本软件提供部分复制和部分备份功能,用于只复制、备份源码文档,可极大地提高案例的切换速度,简化案例库因Android Studio开发平台升级引起的配置同步问题。在进行案例库分发时也极大减小压缩包尺寸。对开发环境有特殊要求的案例可以使用完全复制和完全备份功能,相应案例目录下的所有文件都会进行复制或备份,运行相应案例时无须再进行任何配置设置和变更。为了区分上述两种案例,在运行软件的案例库目录列表区中用红色标识的案例为完整案例;用黑色标识的案例为只含源码文档的案例。

② 双击案例库列表区中目录,该目录背景色改为黄色,同时案例路径显示在下方的"当前选中案例"文本框中,该文本框默认为只读,防止因误触键盘导致对错误目录进行复制或备份。在状态栏右侧显示"当前选中案例顺序号/案例总数量"。

③ "部分复制"按钮。

在执行第②步的操作后单击"部分复制"按钮,软件将"当前选中案例"文本框中显示的路径下src子目录中所有文档(含src下所有子目录和文件)复制到当前开发语言所指定的运行目录中,其他子目录和文件保留不变。在打开的IDE中会刷新显示刚才复制过来的src相关文档。使用本功能前需保证IDE的项目中已经包含项目所需相关辅助文档。

对于Android开发环境频繁升级的情况,建议使用"部分备份"和"部分复制",因为在此模式下并不复制或备份build.gradle之类的配置文件,只要前一个案例设置好相关配置,切换新案例时会沿用之前的配置。特别是Android Studio开发环境的compileSdk、buildTools、Gradle有版本更新时,"部分复制"能平滑地切换到不同的案例,并保证加载的案例按升级后版本正常运行。

④ "完全复制"按钮。

其功能与"部分复制"按钮基本相同,唯一区别是将整个项目文件都复制到运行目录中(前提是案例库中的案例是使用"完全备份"功能保存的案例,这种案例在案例库列表区显示为红色)。

⑤ "部分备份"按钮。

此为"部分复制"的反向操作,将运行目录中的 src 子目录中文档复制到"当前选中案例"文本框所对应目录的 src 子目录中。为防止误操作,单击"部分备份"按钮后会弹出部分备份确认框,单击"是"按钮才开始执行"部分备份"功能。

如果需要新建案例,可双击"当前选中案例"文本框,将其由灰色(只读状态)变成白色,此时可以修改文本框中目录名称。输入将要新建案例的目录路径,单击"部分备份"按钮,系统将自动新建目录并把运行目录中 src 子目录中文档复制到此目录。单击"刷新"按钮,案例库列表区显示变更后的案例库列表。

⑥ "完全备份"按钮。

此为"完全复制"的反向操作,将运行目录中文档全部复制到"当前选中案例"文本框所对应的目录中。为防止误操作,单击"完全备份"按钮后会弹出完全备份确认框,单击"是"按钮才开始执行"完全备份"功能。

如果需要新建案例,可双击"当前选中案例"文本框,将其由灰色(只读状态)变成白色,此时文本框中路径文本可以手工修改。输入将要新建案例的目录路径,单击"完全备份"按钮,系统将自动新建目录并把运行目录中文档全部复制到此目录。单击"刷新"按钮,案例库列表区显示变更后的案例库列表。相应案例目录自动以红色字体标识。

对于案例库中的案例保证至少有一个案例(最好是第一个案例)使用"完全备份"功能。用户在第一次使用时可以使用"完全复制"功能建立运行案例的相同项目环境,后续就可以使用"部分复制"功能运行其他案例。

为防止误操作,"部分复制"和"部分备份"按钮用绿色标识,"完全复制"和"完全备份"按钮用红色标识。

⑦ "显示源码"按钮。

在案例库列表区双击选中案例,单击"显示源码"按钮,软件将显示选中案例目录中的程序源码文件、关键配置文件和 PPT 文件。

此功能是为了方便用户使用文本编辑软件快速查看案例中的源码而无须安装 IDE。源码的显示软件由用户自行定义。Windows 系统自带的记事本、Word、Visual Studio Code、Notepad++、EditPlus 等都可以作为显示源码的文本阅读器,只要用户将相应源码文件扩展名与相应文本编辑软件进行文件关联即可。推荐使用带多文件标签栏的文本编辑软件,便于对一个项目中有多个显示文件的切换浏览和快速关闭。

(11) 软件的状态栏左侧会根据上述不同的操作显示执行结果。状态栏右侧显示"当前选中案例顺序号/案例总数量"信息。

(12) "帮助"菜单显示自带的简单帮助,如图 1-47 所示。

(13) 选择"关于"菜单弹出软件版本信息界面,如图 1-48 所示。

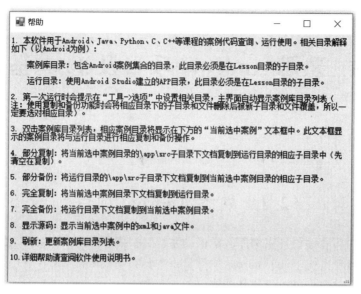

图 1-47 "帮助"菜单

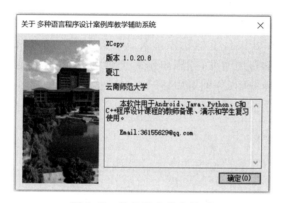

图 1-48 软件版本信息界面

第 2 章　熟悉 Android 项目及 Java 基本概念

2.1　熟悉 Android 项目

视频讲解

通过"多种语言程序设计案例库教学辅助系统"将"熟悉 Android 项目"案例项目使用"完全复制"按钮复制到当前 HelloAndroid 项目目录中。在 Android Studio 中打开 HelloAndroid 项目，接下来按文件类型依次讲述其功能。

2.1.1　Layout

Layout(布局文件)放置在 HelloAndroid\app\src\main\res\layout 目录中，文件名扩展为.xml，用于在 Android 设备上显示图形界面。在导航栏中双击布局文件，在 Android Studio 界面的右侧将显示布局文件，分别单击右上角的 Code、Design 和 Split 三个按钮，界面分别显示为相应的三种视图，分别对应显示布局文件代码、显示图形界面和同时显示文件代码与图形界面。Layout 的 Split 视图如图 2-1 所示。

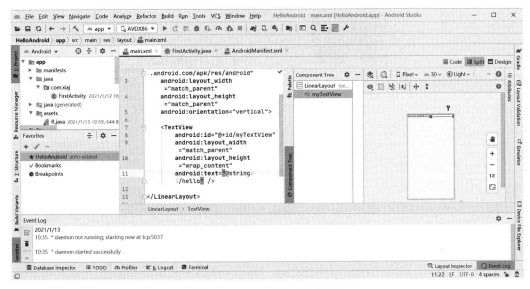

图 2-1　Layout 的 Split 视图

【注】布局文件的文件名只能由小写字母、数字和下画线组成，且不能以数字开头。

Design 视图显示的效果与实际运行的显示效果相似。这里说相似而没有说相同，是因为布局文件是静态的，而实际运行显示效果是综合布局文件、相关配置文件和 Activity 文件相关代码的最终效果。在 Design 视图中可以拖放控件栏的控件到界面中,但目前还无法达到 Visual Studio 的所见即所得效果。在最新版本的 Android Studio 中 Design 视图增加了拖曳修改视图外观尺寸功能,但并未修改布局文件内容,意味着与实际显示效果无关。

Layout 的 Design 视图如图 2-2 所示,在 Design 视图中单击 按钮,在弹出的下拉菜单中可以选择 Design、Blueprint 或者两者同时显示。

在显示系统界面,单击 按钮,在弹出的下拉菜单中选择 Show System UI 菜单,如图 2-3 所示,Design 视图中显示的界面更接近实际运行结果。Design+Blueprint 带 Show System UI 如图 2-4 所示。

图 2-2　Layout 的 Design 视图

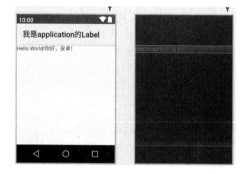

图 2-3　选择 Show System UI 菜单　　图 2-4　Design+Blueprint 带 Show System UI

从 Android Studio 4.0 版本开始提供 Layout Inspector。可以选择 Tools→Layout Inspector 菜单或在 Android Studio 右下角选择 Layout Inspector 选项卡调出 Layout Inspector,如图 2-5 所示。Layout Inspector 可实现对运行中 App 界面的各种布局和控件实现动态查看。

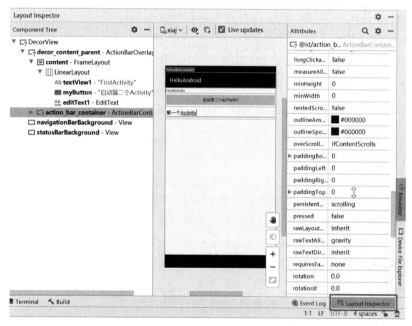

图 2-5　Layout Inspector 选项卡

在实际设计界面的过程中,往往会将 Design 视图和 Code 视图混合使用以完成布局设计。Code 视图在早期 Android Studio 版本中也叫 Text 视图,Code 视图显示的布局文件如下:

```
【main.xml】
01  <? version = "1.0" encoding = "UTF - 8"?>
02  < LinearLayout ns:android = "http://schemas.android.com/apk/res/android"
03      android:layout_width = "match_parent"
04      android:layout_height = "match_parent"
05      android:orientation = "vertical">
06
07      < TextView
08          android:id = "@ + id/textView1"
09          android:layout_width = "match_parent"
10          android:layout_height = "wrap_content"
11          android:text = "@string/hello" />
12  </LinearLayout >
```

第 1 行代表文档开始,其使用的 XML 版本为 1.0,文本编码格式使用 UTF-8。

第 2～12 行定义了一个线性布局容器控件,用 LinearLayout 标签括起。

第 3～5 行分别定义线性布局宽度(与父容器等宽,这里的父容器宽度就是屏幕宽度)、高度(与父容器等高)和内部控件的排列方向(按垂直方向排列内部控件)。

第 7～11 行为 TextView 标签,定义一个 TextView 控件。第 8 行定义 TextView 控件的 id 为 textView1,@ + id 意思是在资源列表中增加一个整型变量,变量名称为 R. id. textView1,具体内容在 2.1.2 节详述。

第 9～10 行指明 TextView 控件的宽度(与父容器等宽,这里的父容器就是第 2 行定义的线性布局)和高度(与文字内容高度保持一致,最大不超过父容器高度)。第 11 行定义 TextView 控件显示的字符串。对字符串的使用有如下两种方式。

(1) 直接使用字符串;

(2) 使用类似@string/hello 方式,其中@string 代表在 HelloAndroid\app\src\main\res\values 路径下的 strings.xml 文件,hello 代表在文件中以下代码中的字符串名:

< string name = "hello"> Hello World! 你好,安卓!</string >

@string/hello 等效于字符串"Hello World! 你好,安卓!"。

整个布局文件 main.xml 定义了一个线性布局,内部放置了一个 TextView 控件。可以在 Code 视图下对布局文件代码直接进行修改,也可以在 Design 视图选中相应控件,然后在属性栏中修改相应属性。Design 视图的属性栏默认只显示常用属性或已定义属性,单击属性栏下方的 All Attributes 可以展开显示选中控件的所有属性。以上两种视图下修改代码是完全等效的,开发人员可根据自己的偏好或者实际情况综合使用以上两种方法。

在 Android Studio 中初次打开布局文件 main.xml,@string/hello 会直接显示为对应的字符串"Hello World! 你好,安卓!",单击此字符串会显示为@string/hello。当鼠标指针放置在@string/hello 上时,自动弹出对应的字符串值,如图 2-6 所示。

图 2-6 字符串资源智能提示

在 strings.xml 文件中增加 string 标签的方式有：

（1）直接在 strings.xml 文件中增加相应 string 标签；

（2）在布局文件中先输入字符串的双引号（半角），将光标置在两个双引号之间，单击代码左侧灯泡形状提示符或按 Alt+Enter 快捷键调出提取字符串资源菜单，如图 2-7 所示。

选择 Extract string resource 菜单，弹出定义资源对话框，如图 2-8 所示。输入字符串资源名称和资源值，单击 OK 按钮，Android Studio 会在 strings.xml 文件中增加一条名为 hello 的字符串资源，同时在布局文件的双引号中自动添加@string/hello 引用。

图 2-7 提取字符串资源菜单

图 2-8 定义资源对话框

【注】 如果布局文件中代码是土黄色背景，同时行首出现黄色警告标识，代表本行代码有更好的编写方式。

2.1.2 Java 文件

布局文件设定了 Android 程序的显示界面，Java 文件则定义了程序的运行逻辑。Java 文件放置的路径为 HelloAndroid\app\src\main\java\com\xiaj，其中的文件代码如下：

```
【FirstActivity.java】
01    package com.xiaj;
02    import android.app.Activity;
03    import android.os.Bundle;
04    public class FirstActivity extends Activity
05    {
06        @Override
07        protected void onCreate(Bundle savedInstanceState)
08        {
09            super.onCreate(savedInstanceState);
10            setContentView(R.layout.main);
11        }
12    }
```

第1行代表当前 Java 所属包为 com.xiaj,关键词 package 所在行必须在代码的最上方。

第2～3行导入后续代码中需要使用到的包,关键词 import 一般用在代码最前面(如果没有定义 package)或者是 package 行后面。import 的包不用强行去记忆或输入完整的名称,例如输入以下代码:

```
Date date = new Date();
```

Android Studio 找不到 Date 类会将之标红,将鼠标指针移动在单词 Date 上时,弹出如图2-9所示的界面,单击 import class 链接或按 Alt+Shift+Enter 快捷键弹出如图2-10所示的提示。提示中有两行,分别代表 java.util.Date 和 java.sql.Date 包,开发人员可根据具体情况自行选择,选择后 Android Studio 自动添加 import 相应包的代码。更快捷的方式是将光标直接放置在单词 Date 上,自动弹出 import 智能提示,如图2-11所示。按下 Alt+Enter 快捷键,如果有多个可以引用包,则如图2-10所示。如果只有唯一可以使用包,则 Android Studio 直接添加 import 相应包的代码。

图 2-9 import 智能提示

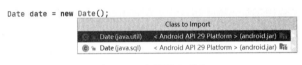

图 2-10 选择导入的包

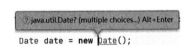

图 2-11 import 光标提示

如果开发人员引入了包,但实际编程代码中并未使用,Android Studio 会将其标识为灰色。在项目完成时可以手工删除灰色代码或者选择 Code→Optimize Imports 菜单(快捷键为 Ctrl+Alt+O)删除没有被使用的包。

第4行代码是建立一个名为 FirstActivity 的类,继承于 Activity(新版本的为 AppCompatActivity,两者差别不大。为方便演示老版本 Android 控件的外观,本书还是沿用 Activity。如果要将案例改为 AppCompatActivity,需要将案例中 AndroidManifest.xml 中 android:theme 属性改为支持 Theme.AppCompat 的主题即可)。Android 有四大组件:Activity、Service(服务)、Content Provider(内容提供)和 BroadcastReceiver(广播接收器)。Activity 是最基本也是最为常用的组件。一个 Activity 通常就是单独显示的一页,可以显示一些控件(可以是动态生成控件或绑定的布局文件),也可以监听事件并对触发事件做出响应。Activity 之间通过 Intent 进行通信。需要强调的是,一个 Java 文件中只能有一个 public 类,而且 public 类的类名称必须与 Java 文件名前缀完全相同(含大小写)。如本例中 public 类为 FirstActivity,其文件名必须为 FirstActivity.java。按照 Java 编程规范,命名采用驼峰命名法,类名首字母大写,类的实例化对象首字母小写。

第6行的@Override 是伪代码,在运行代码时并不执行,所以在程序中不写伪代码也是可以的，但建议加上伪代码@Override。@Override 的作用是提示编译器后面的命令行是重写父类的同名方法。如果开发人员输入的方法名在父类中找不到同名方法则提示出错，Android Studio 会在@Override 下出现红色波浪线。如果没有@Override，且重写的方法名拼写错误时，编译器会认为当前方法是类中新增的方法而非重写父类的方法，不会出现错误提示。在 Android Studio 的文件中只要有一行出现红色波浪线，其文件名也会用红色波浪线标示，方便开发人员快速定位有问题的代码位置。Android Studio 直到目前的版本一直都存在稳定性问题，其中的一个表现就是有时会标示红色波浪线，但程序和配置都没有问题，只要重启 Android Studio 就恢复正常显示。

第 7~11 行是重写父类的 onCreate()方法，也是 Activity 第一次运行的入口（详细内容参见"5.2.3 生存周期"中的案例），其中第 9 行是调用父类的 onCreate()方法。第 10 行是调用布局文件 main.xml 并显示，通过这条命令将 Java 文件的 Activity 与布局文件绑定在一起。Activity 调用 Window，Window 是用来显示图形的。布局文件中的容器、控件都是 View，它们都显示在 Window 上。后续的案例中将会涉及相关概念。setContentView()方法中调用的参数为 R.layout.main 而不是 main 或 main.xml，这与早期的 Android 程序设计有关。在早期的 Android 开发环境（含 Eclipse 和 Android Studio）中编写代码时，会自动生成一个 R.java 的文件，内容如下所示：

```java
【R.java】
/* AUTO-GENERATED FILE. DO NOT MODIFY.
 *
 * This class was automatically generated by the
 * aapt tool from the resource data it found. It
 * should not be modified by hand.
 */

package com.xiaj;

public final class R{
  public static final class drawable{
    public static final int icon = 0x7f010000;
  }
  public static final class id{
    public static final inttextView1 = 0x7f020001;
  }
  public static final class layout{
    public static final int main = 0x7f030000;
  }
  public static final class string{
    public static final int app_name = 0x7f040000;
    public static final int activity_name = 0x7f040001;
    public static final int hello = 0x7f040002;
  }
}
```

Android 项目中增加的任何资源都会被系统自动注册到本文件中。本文件开头的注释指明本文件是由系统自动生成的,不能手工修改。一旦手工修改其中的内容,后续增加的任何资源都不再自动注册,程序也就无法正常运行。这也是后续 Android Studio 版本取消此文件的原因。虽然 R.java 文件已取消,但资源类 R 依然存在。资源类 R 定义为公共最终类,其内部定义了 drawable、id、layout 和 string 内部类(根据开发人员配置的资源可能还会包含其他内部类),分别对应图片、控件、布局和字符串。R.layout.main 是 R 类中内部类 layout 的一个变量,其修饰符为 public static final int,代表是一个公共静态最终 int 型变量,0x7f030000 为变量 main 的值(用十六进制数表示)。因此 R.layout.main 是一个整型值,在资源表中对应 layout 目录下的 main.xml 文件。同样道理,R.id.textView1 对应 main.xml 中的控件 textView1。

【总结】 字符串资源引用方式如下。

Java 代码:getString(R.string.string_name)

XML 代码:@string/string_name

Android Studio 的 Java 文件默认的编码格式为 UTF-8,当导入 GB 2312 或 GBK 的代码时会显示为乱码,单击 Android Studio 窗口右下方状态栏的 UTF-8 按钮,弹出文本编码格式转换提示对话框,如图 2-12 所示。单击 Convert 按钮执行编码格式转换。

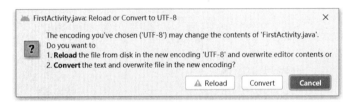

图 2-12 文本编码格式转换提示对话框

【注】 开发中文版 App 可以使用 GB 2312 或 GBK 编码,但这两种编码所包含的汉字有限,可能遇到某些生僻字无法正常显示的问题。推荐使用 UTF-8 编码。

2.1.3 AndroidManifest.xml

本文件是 Android 项目最重要的配置文件,涉及项目的名称、主题、Activity 注册、Service 注册、Receiver 注册、权限注册等。其内容如下:

【AndroidManifest.xml】
```
01  <?xml version = "1.0" encoding = "UTF - 8"?>
02  < manifest xmlns:android = "http://schemas.android.com/apk/res/android"
03       package = "com.xiaj">
04
05       < application
06           android:icon = "@drawable/icon"
07           android:label = "@string/app_name">
08           < activity
09               android:name = "com.xiaj.FirstActivity"
10               android:label = "@string/activity_name"
```

```
11                    android:theme = "@android:style/Theme.Holo.Light">
12                    <!-- 改变主题 -->
13                    <intent-filter>
14                        <action android:name = "android.intent.action.MAIN" />
15                        <category android:name = "android.intent.category.LAUNCHER" />
16                    </intent-filter>
17             </activity>
18
19      </application>
20 </manifest>
```

第 5~19 行为 application 标签，其中第 6 行定义项目的图标，第 7 行定义项目的标题名称。

第 8~17 行是在 application 中注册一个 Activity，其名称为 com.xiaj 包下的 FirstActivity，第 10 行定义 Activity 标签名称，第 11 行定义 Activity 的主题，第 13~16 行指明当前 Activity 为项目启动时优先运行的 Activity(有多个 Activity 时解决哪一个 Activity 会首先启动)。后续案例中涉及的服务、广播或者其他相关权限也必须要在此文件中注册。当程序调用未注册资源时，相关功能将无法正常使用或出现闪退现象。

文件中定义了两个 label 属性：一个是第 7 行 application 标签中的 label 属性，指向变量 app_name；另一个是第 10 行 activity 标签中的 label 属性，指向变量 activity_name。strings.xml 文件内容如下：

```
01 <?xml version = "1.0" encoding = "UTF-8"?>
02 <resources>
03     <string name = "hello">Hello World!你好,安卓!</string>
04     <string name = "app_name">我是 Application 的 Label</string>
05     <string name = "activity_name">我是 Activity 的 Label</string>
06 </resources>
```

第 4 行指明 app_name 对应字符串"我是 Application 的 Label"，第 5 行 activity_name 对应字符串"我是 Activity 的 Label"。

标题栏显示 activity 标签中的 label 属性，运行结果如图 2-13 所示。Android 设备上 App 的名称为"我是 Activity 的 Label"。

把 AndroidManifest.xml 中第 10 行 activity 标签中的 label 属性删除，再次运行程序显示 application 标签中的 label 属性，如图 2-14 所示，标题栏显示的是 application 标签中的 label 属性。Android 设备上 App 的名称为"我是 Application 的 Label"。

由此可以看出，当 AndroidManifest.xml 中的 application 标签和 activity 标签中都定义了 label 时，优先选择 activity 标签中的 label 属性。如果项目中包含一个以上的 Activity，每个 Activity 的标题栏显示各自 activity 标签中定义的 label 属性，App 的名称也由优先启动 Activity 对应的 label 属性确定。当 activity 标签中未定义 label 属性时，App 的名称由 application 标签中的 label 属性确定。

图 2-13　显示 activity 标签中的 label 属性　　　图 2-14　显示 application 标签中的 label 属性

也可以在 Java 文件(更准确的描述是用户定义的 Activity 继承类，如 FirstActivity 类，后续描述中用 Java 文件指代)中使用如下命令：

```
setTitle("在 Java 代码设置的 Activity 标题");
```

该命令用于动态设置 Activity 标题栏显示字符，优先级高于 AndroidManifest.xml 中 activity 标签中的 label 属性，但该命令不修改 App 名称。

【注】　编程中默认遵循作用域越小，优先级越高的原则。

2.2　调 试 输 出

除了使用 Debug 进行中断调试以外，还可以采用以下两种方式输出中间运行结果。
(1) 使用 Java 语言的 System.out.println()方法输出结果，如插入以下代码：

```
System.out.println("变量 i = " + i);
```

在 Logcat 窗口中输出以下信息：

```
2021 - 01 - 15 16:52:47.854 12807 - 12807/com.xiaj I/System.out:变量 i = 1
```

在 Logcat 窗口会显示大量的调试信息，可以在过滤文本框中输入上述信息的任意字符串作为过滤关键字来筛选输出结果。过滤显示 System.out.println 输出的 Logcat，如图 2-15 所示

(2) 使用 Log 类输出结果。
Android 中单独提供了 Log 类处理信息的输出。Log 类中有 d()、i()、w()、e()、v()方

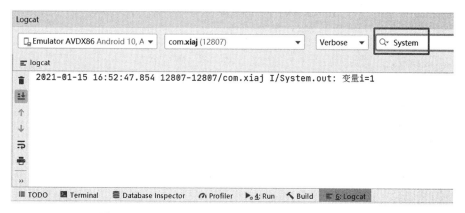

图 2-15　过滤显示 System.out.println 输出的 Logcat

法,分别对应 DEBUG、INFO、WARN、ERROR 和 VERBOSE 日志信息,输出的信息颜色稍有不同。常用方法的语法格式为:

```
public static int d(String tag, String msg)
public static int i(String tag, String msg)
public static int w(String tag, String msg)
public static int e(String tag, String msg)
public static int v(String tag, String msg)
```

其中,参数 tag 用于在输出信息中添加一个字符串标记,自定义标记可以作为输出结果的过滤器关键字使用;参数 msg 是要输出的信息。例如输入如下代码:

```
Log.d("xj", "变量j=" + j);
Log.i("xj", "变量j=" + j);
Log.w("xj", "变量j=" + j);
Log.e("xj", "变量j=" + j);
Log.v("xj", "变量j=" + j);
```

Log 输出的 Logcat 如图 2-16 所示。在右上方文本框输入的字符串"xj"作为过滤关键字,只有包含关键字的输出行信息才显示。Log.w 输出信息为红色,Log.e 输出信息为蓝色,其余为黑色。

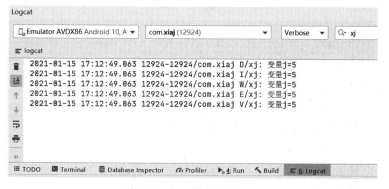

图 2-16　Log 输出的 Logcat

单击 Logcat 窗口左下角的 Logcat Header 按钮 ⚙，配置 Logcat 输出信息内容，如图 2-17 所示。

图 2-17　配置 Logcat 输出信息内容

如果取消所有选中的复选框，则显示简化输出信息的 Logcat，如图 2-18 所示。

图 2-18　简化输出信息的 Logcat

感兴趣的读者也可以在命令提示符或者 Android Studio 下方的 Terminal 窗口进入 Sdk\platform-tools 目录（最好将目录加入到 Windows 环境变量 PATH 中），输入 adb logcat，可看到与 Android Studio 中 Logcat 窗口相同的输出信息。

2.3　类 和 对 象

从本节开始，将通过案例来了解 Android 开发中 Java 的相关基本概念。面向对象编程语言最重要的一个概念就是类和对象。类是组成 Java 程序的基本要素，它封装了属性和方法。通过"类名 引用变量名 = new 构造方法();"的方式将类实例化为对象。标识符（含类名、方法名、变量名等）的命名规范如下：

(1) 组成类名的各单词首字母大写，剩余字母小写；

(2) 变量名、方法名首字母小写，其余单词首字母大写，剩余字母小写；

(3) 标识符只能由字母、数字、下画线、$ 符号组成；

(4) 不能以数字开头；

(5) 标识符不能使用 Java 和 Android 的关键字。

从本案例起,只将涉及的代码和关键配置文件内容列出,其余代码和配置文件可参见案例库。

【FirstActivity.java】
```
01  public class FirstActivity extends Activity
02  {
03      @Override
04      public void onCreate(Bundle savedInstanceState)
05      {
06          super.onCreate(savedInstanceState);
07          setContentView(R.layout.main);
08
09          student("方法");   //这是调用 FirstActivity 类中的 student()方法
10          Student student1 = new Student();    //这是对 Student 类的实例化,不带参数
11          Student student2 = new Student("李四");//这是对 Student 类的实例化,带参数
12          Log.i("xj","通过调用类实例 student2 中的 getName()方法: " + student2.getName() + ",
                student1.count = " + student1.count + ",Student.count = " + Student.count);
13      }
14
15      String student(String name)
16      {
17          Log.i("xj","姓名 = " + name);
18          return "姓名 = " + name;
19      }
20  }
21
22  class Student
23  {
24      String name = "";           //成员变量中的实例变量
25      static int count;           //成员变量中的类变量
26
27      Student()
28      {
29          count++;
30          Log.i("xj","name = 匿名,count = " + count);
31      }
32
33      Student(String name)
34      {
35          count++;
36          Log.i("xj","姓名 = " + name + ",count = " + count);
37          this.name = name;        //分清楚两个 name 的区别
38      }
39
40      String getName()
41      {
42          return "姓名是" + name;
43      }
44  }
```

在 FirstActivity.java 中定义了两个类：FirstActivity 和 Student，分别对应第 1~20 行和第 22~44 行。两个类定义在一个 Java 文件中或者分别放置在两个 Java 文件中的效果是一样的。Student 类也可以放置在第 19 行与 20 行之间，此时 Student 类变成了 FirstActivity 类的内部类。回顾前面讲述的内容，FirstActivity 是 FirstActivity.java 文件中唯一用关键字 public 修饰的类，类名必须和文件名前缀完全一致。FirstActivity 类的运行入口为 onCreate()方法。

程序运行到第 9 行，调用 FirstActivity 类中的 student()方法，为便于区别于类，方法名首字母小写。括号内是 student()方法的实参。程序运行到此行会调用第 15~19 行的 student()方法，在 Logcat 中输出"姓名=方法"。

程序运行完 student()方法后返回到第 10 行，声明并创建一个名为 student1 的对象实例，关键字 new 后的 Student()是 Student 类的无参构造方法。构造方法名必须与类名同名，因此构造方法的首字母也大写。程序转到第 27~31 行运行 Student()无参构造方法。

【注】 构造方法与普通方法的区别如下。

（1）构造方法的首字母大写，普通方法首字母小写。

（2）构造方法不能指定返回类型，默认返回类的实例。普通方法需要指明方法返回类型，如果没有返回类型，方法名前要用关键字 void 修饰。

（3）如果类中没有定义构造方法，则隐含调用父类的无参构造方法；如果父类没有定义无参构造方法，则依次向上寻找，直至调用 Object 类的无参构造方法。如果类中的普通方法没有定义（含重载），则隐含调用父类的同名方法；如果父类中也没有定义，则程序出错。

（4）子类构造方法的第一条命令前隐含一条 super()，调用父类的无参构造方法。子类普通方法中则没有 super()，如果要调用父类中的同名普通方法使用 super.方法名。

程序返回第 11 行，声明并创建一个名为 student2 的对象实例，此时调用的是带参数的 Student()构造方法，对应第 33~38 行。在 Student 类中同时存在无参构造方法和有参构造方法，将其称为重载，具体内容在案例"方法重载"中详解。

最后程序执行到第 12 行，student2.getName()调用实例变量 student2 中的 getName()方法。这是调用类中成员变量的常用方式。在本案例中要搞清楚变量关系图，如图 2-19 所示。

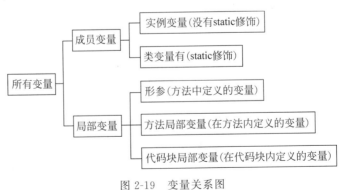

图 2-19 变量关系图

实例变量是与类的实例共存亡的变量，它随着类实例化而建立，随着类实例的消亡而删除。第 24 行中的 name 属于成员变量中的实例变量，类实例 student1 和 student2 中的

name 有不同的存储地址,属于两个不同的变量。变量的使用要遵循作用域的就近规则,如第 42 行的 name 与第 24 行声明的 name 属于同一个作用域,是同一个变量。第 37 行等号右边的 name 对应第 33 行的局部变量 name,由于此局部变量 name 与第 24 行的实例变量 name 同名,为区分两者,实例变量 name 前要加关键字 this 以示区别,如 this.name。

有 static 修饰的成员变量为静态变量,也叫类变量。顾名思义,类变量是与类共存亡的变量。对于同一个类的所有实例,类变量共享同一个存储地址,即所有类实例中的类变量是同一个。如第 25 行类变量 count 对类实例 student1 和 student2 是共享的,各自执行加 1 操作都是对共享的类变量 count 依次执行加 1 操作。定义类时类变量就建立,类变量可以不用将类实例化以后再调用。如第 12 行中 student1.count 是调用类实例 student1 中的 count,Student.count 是直接调用 Student 类中的类变量 count,注意两者首字母 s 的大小写差异。虽说两者调用的都是同一个变量,但概念是有差异的。

【注】 count 变量没有赋初值就执行 ++ 操作而没有报错,原因是成员变量如果是基础数据类型,其默认初值为 0 或 0.0,包装类数据类型的成员变量默认初值为 null。如果是局部变量则必须赋初值后才能使用。

程序运行结果如下:

```
I:姓名 = 方法
I: name = 匿名,count = 1
I:姓名 = 李四,count = 2
I:通过调用类实例 student2 中的 getName()方法:姓名是李四,student1.count = 2,Student.count = 2
```

2.4　extends

通过本案例了解继承、父类、子类以及直接父类(子类)和间接父类(子类)的概念。

视频讲解

```
【FirstActivity.java】
01    public class FirstActivity extends Activity
02    {
03        @Override
04        public void onCreate(Bundle savedInstanceState)
05        {
06            super.onCreate(savedInstanceState);
07            setContentView(R.layout.main);
08    
09            //1.类正常实例化
10            Animal animal = new Animal();
11            animal.say();
12            animal.getName();
13            Animal.getName();
14    
15            Dog dog = new Dog();
16            dog.say();
17            dog.getName();
18    
```

```java
19              Cat cat = new Cat();
20              cat.say();
21              cat.getName();
22
23              (new PersianCat()).say();
24              (new PersianCat()).getName();
25              //2.父类引用指向Dog子类的具体实现
26              //Animal newAnimal = new Dog();
27              //newAnimal.say();
28              //newAnimal.getName();
29
30              //Cat cat2 = new Animal();
31      }
32 }
33
34 class Animal
35 {
36      void say()
37      {
38          Log.i("xj","调用Animal的方法say():此时无声胜有声");
39      }
40
41      static void getName()
42      {
43          Log.i("xj","调用Animal的方法getName()");
44      }
45 }
46
47 class Dog extends Animal
48 {
49      @Override
50      void say()
51      {
52          //super.say();
53          Log.i("xj","调用Dog的方法say():汪汪汪");
54      }
55
56      static void getName()
57      {
58          Log.i("xj","调用Dog的方法getName()");
59      }
60 }
61
62 class Cat extends Animal
63 {
64      void say()
65      {
66          Log.i("xj","调用Cat的方法say():喵喵喵");
67      }
68
```

```
69          static void getName()
70          {
71                  Log.i("xj", "调用 Cat 的方法 getName()");
72          }
73      }
74
75      class PersianCat extends Cat
76      {
77          void say()
78          {
79                  Log.i("xj", "调用 PersianCat 的方法 say(): 波斯猫喵喵喵");
80          }
81
82          static void getName()
83          {
84                  Log.i("xj", "调用 PersianCat 的方法 getName()");
85          }
86      }
```

第34~45行定义了Animal类,内部定义了say()成员方法和getName()静态成员方法(类方法)。

第47行使用关键词extends指明Dog类是Animal类的子类;同样,在第62行指明Cat类是Animal类的子类。第75行指明PersianCat类是Cat类的子类。由此可知,Animal类是Cat类的父类或直接父类,Cat类是PersianCat类的父类或直接父类,Animal类是PersianCat类的间接父类。反之,Cat类是Animal类的子类或直接子类,PersianCat类是Cat类的子类或直接子类,PersianCat类是Animal类的间接子类。

第10行通过Animal()构造方法实例化一个Animal类并赋给实例变量animal。

第11行调用实例变量animal中的say()方法。

以上两条命令也可以采用第23行的方式合二为一。

第13行是使用类名直接调用getName()方法。与类变量相似,类方法也可以不用初始化就直接调用。

程序运行结果如下:

```
I:调用 Animal 的方法 say(): 此时无声胜有声
I:调用 Animal 的方法 getName()
I:调用 Animal 的方法 getName()
I:调用 Dog 的方法 say(): 汪汪汪
I:调用 Dog 的方法 getName()
I:调用 Cat 的方法 say(): 喵喵喵
I:调用 Cat 的方法 getName()
I:调用 PersianCat 的方法 say(): 波斯猫喵喵喵
I:调用 PersianCat 的方法 getName()
```

如果删除第77~85行的代码(PersianCat类中没有定义say()和getName()方法),程序运行结果如下:

```
I: 调用 Animal 的方法 say(): 此时无声胜有声
I: 调用 Animal 的方法 getName()
I: 调用 Animal 的方法 getName()
I: 调用 Dog 的方法 say(): 汪汪汪
I: 调用 Dog 的方法 getName()
I: 调用 Cat 的方法 say(): 喵喵喵
I: 调用 Cat 的方法 getName()
I: 调用 Cat 的方法 say(): 喵喵喵
I: 调用 Cat 的方法 getName()
```

可以看到最后两行的结果发生了变化,说明如果子类实例调用的方法在子类没有定义但在父类中有定义时会自动调用父类中的同名方法。

把第 10~24 行删除,去掉第 26~28 行的注释,程序运行结果如下:

```
I: 调用 Dog 的方法 say(): 汪汪汪
I: 调用 Animal 的方法 getName()
```

这是将子类的实例化对象赋给父类实例变量,其方法调用顺序遵循:

(1) 如果调用的是实例方法(不含 static 修饰的成员方法),优先调用子类中的实例方法,子类中没有的实例方法才调用父类中的同名实例方法。

(2) 如果调用的是类方法(含 static 修饰的成员方法)则还是优先调用各自类中的同名类方法(静态变量和静态方法是在定义类时就建立的,与是否实例化无关)。

因此第 27 行的 say()方法属于实例方法,调用的是 Dog 类中定义的 say()方法。

第 28 行的 getName()方法属于类方法,调用的是 Animal 类中定义的 getName()方法。

当输入代码"**Cat cat＝new Animal()；**"时,Android Studio 会在代码下方标上红色波浪线,说明这行代码有错误。这行代码的逻辑是:动物是猫。这明显是错误的,因为动物也可能是狗。反过来"**Animal animal＝new Cat()；**"就没有问题,这行代码的逻辑是:猫是动物。

如果子类方法中要调用父类的方法,可使用第 52 行"super.方法名"的方式调用父类中的方法。如果是构造方法则默认有 super()方法,具体参见"2.6 super"中的案例。

2.5　implements

视频讲解

Java 不支持 extends 的多重继承,但是可以通过 implements 接口实现比多重继承更强的功能。一个类可以有多个接口,接口间用逗号分隔。

【注】 implements 接口类的属性和方法默认修饰关键字:

(1) 接口中默认变量的修饰是 public static final。

(2) 接口中默认方法的修饰是 public abstract。

【FirstActivity.java】
```
01    public class FirstActivity extends Activity
02    {
03        @Override
04        public void onCreate(Bundle savedInstanceState)
```

```
05          {
06                  super.onCreate(savedInstanceState);
07                  setContentView(R.layout.main);
08
09                  Dog dog = new Dog();
10                  dog.say();
11                  Log.i("xj", "------------------------------------");
12
13                  Cat cat = new Cat();
14                  cat.say();
15                  Log.i("xj", "------------------------------------");
16
17                  //父类引用指向Dog子类的具体实现
18                  Animal animal = new Dog();
19                  animal.say();
20                  animal = new Cat();
21                  animal.say();
22                  Log.i("xj", "------------------------------------");
23
24                  Animal animal1 = new Animal()
25                  {
26                          @Override
27                          public void say()
28                          {
29                                  Log.i("xj","调用匿名类中的方法say():汪～～");
30                          }
31                  };
32                  animal1.say();
33
34                  //以上代码等效于
35                  (new Animal()
36                  {
37                          @Override
38                          public void say()
39                          {
40                                  Log.i("xj","调用匿名类中的方法say():汪");
41                          }
42                  }).say();
43          }
44  }
45
46  interface Animal
47  {
48          String name = "动物世界";
49
50          void say();
51  }
52
53  class Dog implements Animal
54  {
```

```
55          static String name = "Dog";
56          @Override
57          public void say()    //需重写父类方法 say(),前面必须要有关键字 public
58          {
59              Log.i("xj", "name = " + name + ", Animal.name = " + Animal.name + ", Dog.
                    name = " + Dog.name);
60              Log.i("xj", "调用 Dog 的方法 say():汪汪汪");
61          }
62      }
63
64      class Cat implements Animal
65      {
66          @Override
67          public void say()
68          {
69              Log.i("xj", "name = " + name + ", Animal.name = " + Animal.name + ", Cat.
                    name = " + Cat.name);
70              Log.i("xj", "调用 Cat 的方法 say(): 喵喵喵");
71          }
72      }
```

本案例是将 2.4 节案例中的 extends 换为 implements,重点关注 implements 接口中变量和方法所具有的默认关键字修饰。

第 48 行的变量 name 默认由 public static final 修饰,实现接口的 Cat 类中第 69 行可以直接使用变量 Cat.name,Cat 类中并没有定义变量 name,实际调用的是第 48 行接口 Animal 的静态变量 name,所以 Animal.name 与 Cat.name 对应的值相同。

第 55 行 Dog 类中定义的 name 覆盖接口 Animal 中的 name,所以 Animal.name 与 Dog.name 对应的值不同。

由于接口不能直接实例化使用,因此第 24~31 行是将接口派生的匿名类(需重写实现其中的抽象方法,第 27~30 行实现了接口中的 say()方法)实例化后传给 animal1。当第 24 行代码输入到 new Animal 时,Android Studio 会弹出智能提示框,按下 Tab 键或 Enter 键,Android Studio 会自动生成第 25~31 行的代码框架。开发人员只要填写 say()方法的具体实现代码并补上第 31 行的分号就行了。

第 35~42 行代码等效于第 24~32 行代码。

程序运行结果如下:

```
I: name = Dog, Animal.name = 动物世界, Dog.name = Dog
I: 调用 Dog 的方法 say():汪汪汪
I: --------------------------------
I: name = 动物世界, Animal.name = 动物世界, Cat.name = 动物世界
I: 调用 Cat 的方法 say(): 喵喵喵
I: --------------------------------
I: name = Dog, Animal.name = 动物世界, Dog.name = Dog
I: 调用 Dog 的方法 say():汪汪汪
I: name = 动物世界, Animal.name = 动物世界, Cat.name = 动物世界
I: 调用 Cat 的方法 say(): 喵喵喵
```

```
I:------------------------------------
I:调用匿名类中的方法 say():汪~~
I:调用匿名类中的方法 say():汪
```

在 Android 应用软件开发中,监听器就是采用接口方式响应回调,对监听器接口的使用方式与上述代码相同。

抽象类与接口有很多相似之处,但原理和使用方式还是有差异,抽象类与接口的对比参见表 2-1。

表 2-1 抽象类与接口的对比

对比项	抽象类	接口
变量	与普通类中的变量没有区别	默认由 public static final 修饰
方法	方法可以有代码,如果是抽象方法则不能有具体代码,由继承的子类提供抽象类中所有声明的抽象方法的实现	方法默认的修饰是 public abstract,不能有具体的代码。由实现接口的类提供接口中所有声明方法的实现
构造方法	抽象类可以有构造方法	接口不能有构造方法
与普通类的区别	除了不能实例化抽象类之外,和普通类没有区别	接口是完全不同的类型
多继承	抽象方法可以继承一个类或实现多个接口	接口可以继承一个类或实现多个接口

【提问】 abstract 和 final 可以同时修饰一个类吗?

abstract 修饰的类为抽象类,表示的是一个抽象概念,不能被实例化为对象。抽象类通过继承实现具体应用,其中的抽象方法在子类中必须要实现。这就意味着抽象类必须由继承的子类具体实现,而 final 类型的类不能变更,因此两个关键字不能同时修饰一个类。

2.6 super

通过本案例要掌握 super 的使用方式及带参数 super()方法、不带参数 super()方法和隐含 super()方法的运行变化。super()的意思是当前类的父类。

【FirstActivity.java】
```
01  public class FirstActivity extends Activity
02  {
03      @Override
04      public void onCreate(Bundle savedInstanceState)
05      {
06          super.onCreate(savedInstanceState);
07          setContentView(R.layout.main);
08          Son son = new Son();
09      }
10  }
11
12  class Father
13  {
14      int age = 30;
```

```
15
16          Father()
17          {
18              Log.i("xj", "Father ():age = " + age + "岁");
19          }
20
21          Father(String name)
22          {
23              Log.i("xj", "Father(name):name = " + name + "  age = " + age + "岁");
24          }
25      }
26
27      class Son extends Father
28      {
29          int age = 10;
30
31          Son()
32          {
33
34              Log.i("xj", "Son (): age = " + age + "岁");
35              Log.i("xj", "Son:super.age = " + super.age + "岁");
36          }
37
38          Son(String name)
39          {
40
41              Log.i("xj", "Son(name):name = " + name + "  age = " + age + "岁");
42              Log.i("xj", "Son(name):super. age = " + super.age + "岁");
43          }
44      }
```

程序运行到第 8 行时调用 Son 类的无参数构造方法,转到第 31 行运行 Son()无参数构造方法。Java 中规定,子类的构造方法中隐含一条 super 命令来调用父类的同类型(指参数类型和数量一致或参数类型可进行隐式转换的)构造方法。如果手工添加 super()方法,必须是构造方法的第 1 条命令,且不能用在非构造方法中。所以程序运行到第 31 行后会因隐含的"**super**();"命令而自动转至第 16 行执行父类的无参数构造方法并输出 Father 中的变量 age,之后又转回第 34 行输出 Son 中的变量 age。第 35 行的 super.age 是调用 Son 的父类 Father 中的变量 age。程序运行结果如下:

```
I: Father (): age = 30 岁
I: Son (): age = 10 岁
I: Son ():super.age = 30 岁
```

如果在第 33 行加入"**super**("张三");"代码,当程序运行到此行时发现有带参数的 super()方法,就调用带参数的父类构造方法并忽略隐含的无参数 super()方法,转而执行第 21 行,将实参字符串"张三"传递给形参 name。执行完父类带参数构造方法之后,再转回第 34 行继续执行。程序运行结果如下:

```
I: Father(name):name = 张三    age = 30 岁
I: Son ( ): age = 10 岁
I: Son ( ):super.age = 30 岁
```

返回到最初的代码,将第 8 行的命令换成"**Son son ＝ new Son**("张娃");",当运行到此行代码时转第 38 行执行带参数的构造方法。此时构造方法的第一行不是 super()方法,就调用隐含的不带参数的 super()方法,因此转到第 16 行执行父类不带参数构造方法。最后再转回第 41 行继续执行。程序运行结果如下:

```
I: Father ( ):age = 30 岁
I: Son(name):name = 张娃    age = 10 岁
I: Son(i):super. age = 30 岁
```

【提问】 第 35 行有 super.age,可以改成 super.name 吗?
age 在父类 Father 中是成员变量,子类 Son 可以使用 super.age 来调用父类中的成员变量 age。name 在父类 Father 中属于局部变量,是带参数构造方法中的形参,子类 Son 无法通过 super.name 来调用父类 Father 中的局部变量。"super."只能用于调用父类成员变量和成员方法(不包含构造方法)。

【提问】 如果 Son 类中没有定义任何构造方法,当运行 new Son()时会如何?
会自动调用 Father 类中的无参构造方法。

【提问】 将 Father 类中的无参数构造方法注释掉会怎样?
Son 类的构造方法如果没有指定 super()方法,默认调用父类 Father 类不带参数的构造方法,而 Father 类此时无参数构造方法未定义(已被注释),Android Studio 会在 Son 构造方法下用红色波浪线标识,提示在 com. xiaj. Father 中没有默认构造方法可供使用。所以设计类通常要提供无参构造方法。

【提问】 如果 Father 类中没有定义任何构造方法,当运行 new Son()时会如何?
运行 new Son()时会默认调用 super(),即 Father 的无参构造方法。虽然 Father 中没有定义构造方法,但默认隐含调用 Father 的父类 Object 的无参构造方法。

2.7　equals 与恒等号(＝＝)

视频讲解

Java 中对象或值的比较有恒等号"＝＝"和 equals 两种方式。初学者对于这两个概念存在认识误区。

(1) 从 C 语言转过来的开发人员习惯使用恒等号比较两个变量的值是否相等。
(2) 很多 Java 的书籍中提到 equals 比较的是对象的值,恒等号比较的是对象的地址。

以上两种说法有些片面,通过下面的案例对 equals 和恒等号有一个正确和全面的认识。

【FirstActivity.java】
```
01    public class FirstActivity extends Activity
02    {
03        @Override
```

```
04      public void onCreate(Bundle savedInstanceState)
05      {
06              super.onCreate(savedInstanceState);
07              setContentView(R.layout.main);
08
09              Object object1 = new Object();
10              Object object2 = new Object();
11              Log.i("xj", " Object object1 = new Object();\nObject object2 = new Object();");
12              Log.i("xj", "object1 == object2 is " + (object1 == object2));
13              Log.i("xj", "object1.equals(object2) is " + (object1.equals(object2)));
14
15              Log.i("xj", "------------------------------");
16              object1 = object2;
17              Log.i("xj", "执行 object1 = object2 后比较");
18              Log.i("xj", "object1 == object2 is " + (object1 == object2));
19              Log.i("xj", "object1.equals(object2) is " + (object1.equals(object2)));
20
21              Log.i("xj", "------------------------------");
22              String str1 = new String("a");
23              String str2 = new String("a");
24              Log.i("xj", " String str1 = new String(\"a\");\nString str2 = new String(\"a\");");
25              stringCompare(str1, str2);
26
27              Log.i("xj", "------------------------------");
28              String str3 = "a";
29              String str4 = "a";
30              Log.i("xj", " String str3 = \"a\";\nString str4 = \"a\";");
31              stringCompare(str3, str4);
32
33              Log.i("xj", "------------------------------");
34              int i = 1;
35              int j = new Integer(1);
36              Log.i("xj", " int i = 1;\nint j = new Integer(1);");
37              Log.i("xj", "基本数据类型 int i == j is " + (i == j));
38
39              Log.i("xj", "------------------------------");
40              Integer m = new Integer(1);
41              Integer n = new Integer(1);
42              Integer x = Integer.valueOf(1);
43              Integer y = Integer.valueOf(1);
44              Log.i("xj", " Integer m = new Integer(1);\nInteger n = new Integer(1);\nInteger x = Integer.valueOf(1);\nInteger y = Integer.valueOf(1);");
45              Log.i("xj", "Integer 对象 m == n is " + (m == n));
46              Log.i("xj", "Integer 对象 m equals n is " + (m.equals(n)));
47              Log.i("xj", "Integer 对象 m == x is " + (m == x));
48              Log.i("xj", "Integer 对象 m equals x is " + (m.equals(x)));
49              Log.i("xj", "Integer 对象 x == y is " + (x == y));
50              Log.i("xj", "Integer 对象 x equals y is " + (x.equals(y)));
51
```

```
52              Log.i("xj", "------------------------------");
53              Integer a = 100;
54              Integer b = 100;
55              Integer c = 200;
56              Integer d = 200;
57              Log.i("xj", " Integer a = 100;\nInteger b = 100;\nInteger c = 200;\nInteger d = 200;");
58              Log.i("xj", "a == b is " + (a == b));
59              Log.i("xj", "c == d is " + (c == d));
60
61              Log.i("xj", "------------------------------");
62              Integer x1 = Integer.valueOf(100);
63              Integer y1 = Integer.valueOf(100);
64              Integer x2 = Integer.valueOf(200);
65              Integer y2 = Integer.valueOf(200);
66              Log.i("xj", " Integer x1 = Integer.valueOf(100);\nInteger y1 = Integer.valueOf(100);\nInteger x2 = Integer.valueOf(200);\nInteger y2 = Integer.valueOf(200);");
67              Log.i("xj", "Integer 对象 x1 == y1 is " + (x1 == y1));
68              Log.i("xj", "Integer 对象 x1 equals y1 is " + (x1.equals(y1)));
69              Log.i("xj", "Integer 对象 x2 == y2 is " + (x2 == y2));
70              Log.i("xj", "Integer 对象 x2 equals y2 is " + (x2.equals(y2)));
71          }
72
73          /**
74           * 将两个字符串比较 == 和 equals 的结果
75           *
76           * @param str1 字符串1
77           * @param str2 字符串2
78           */
79          static void stringCompare(String str1, String str2)
80          {
81              //if (str1 == str2)
82                  //System.out.println("两字符串 == true");
83              //else
84                  //System.out.println("两字符串 == false");
85              //if (str1.equals(str2))
86                  //System.out.println("两字符串 equals true");
87              //else
88                  //System.out.println("两字符串 equals false");
89
90              Log.i("xj", "两字符串 == " + (str1 == str2));   //== 比较的两字符串变量
                                                                //要用括号括起来
91              Log.i("xj", "两字符串 equals " + str1.equals(str2));
92          }
93      }
```

第79～92行自定义 stringCompare() 方法实现两个字符串的 equals 与恒等比较。很多开发人员会采用第81～88行的代码实现两个变量的比较,可以采用第90～91行的代码来简化程序设计、提高代码运行效率和可读性。这只是取消 if-else 代码结构的一种方案,后续案例中会演示其他替代方案。程序运行结果如下:

```
01  I:  Object object1 = new Object();
02      Object object2 = new Object();
03  I: object1 == object2 is false
04  I: object1.equals(object2) is false
05  I: ------------------------------
06  I: 执行 object1 = object2 后比较
07  I: object1 == object2 is true
08  I: object1.equals(object2) is true
09  I: ------------------------------
10  I:  String str1 = new String("a");
11      String str2 = new String("a");
12  I: 两字符串 == false
13  I: 两字符串 equals true
14  I: ------------------------------
15  I:  String str3 = "a";
16      String str4 = "a";
17  I: 两字符串 == true
18  I: 两字符串 equals true
19  I: ------------------------------
20  I:  int i = 1;
21      int j = new Integer(1);
22  I: 基本数据类型 int i == j is true
23  I: ------------------------------
24  I:  Integer m = new Integer(1);
25      Integer n = new Integer(1);
26      Integer x = Integer.valueOf(1);
27      Integer y = Integer.valueOf(1);
28  I: Integer 对象 m == n is false
29  I: Integer 对象 m equals n is true
30  I: Integer 对象 m == x is false
31  I: Integer 对象 m equals x is true
32  I: Integer 对象 x == y is true
33  I: Integer 对象 x equals y is true
34  I: ------------------------------
35  I:  Integer a = 100;
36      Integer b = 100;
37      Integer c = 200;
38      Integer d = 200;
39  I: a == b is true
40  I: c == d is false
41  I: ------------------------------
42  I:  Integer x1 = Integer.valueOf(100);
43      Integer y1 = Integer.valueOf(100);
44      Integer x2 = Integer.valueOf(200);
45      Integer y2 = Integer.valueOf(200);
46  I: Integer 对象 x1 == y1 is true
47  I: Integer 对象 x1 equals y1 is true
48  I: Integer 对象 x2 == y2 is false
49  I: Integer 对象 x2 equals y2 is true
```

为了便于查看运行结果,将变量赋值代码转换为字符串先输出显示,然后列出＝＝或 equals 的比较结果。对比代码和运行结果是否有些令人意外?

【注 1】 恒等号(＝＝)返回 true 的条件是恒等号两边的变量或值使用相同的存储地址。使用时需注意:

(1) 用于比较两个基本数据类型变量是否相等,等同于比较两个基本数据类型变量的值是否相等。

(2) 用于比较两个包装类变量是否相等,即判断这两个包装类变量是否使用相同的存储地址,也就是判断是否是同一个包装类实例的不同名称。如果两个包装类变量的值相等,但地址不一样,恒等号比较的结果为 false,只有地址一样比较的结果才为 true。

【注 2】 equals 使用方法:x.equals(y),比较对象变量 x 的值是否等于对象变量 y 的值,返回 true 的条件是两个变量的值相等(而存储地址可能不相同),Object 类型数据变量除外。Ctrl+鼠标左键单击 equals,Android Studio 会自动打开 Object.java 文件并自动转到如下代码行:

```
public boolean equals(Object obj) {
        return (this == obj);
    }
```

从源码中可以看出在 Object 类中,equals()方法是采用恒等号运算进行比较。

【总结】 在包装类对象的比较中,恒等号"＝＝"返回 true,equals()也返回 true(没有定义或重写了 equals()方法的除外)。

上述【注 1】和【注 2】能解释本案例的大多数代码的运行结果,但还是有少量代码的结果与预期不符。下面对运行结果逐一加以分析。

Object 类是所有 Java 类的始祖类,其中的 equals()方法是采用＝＝运算进行比较(其他从 Object 派生的类一般都会重写 equals()方法),而＝＝是比较两个对象的存储地址。一般使用 new 关键词生成的对象实例都会分配不同的存储地址,所以输出结果的第 3 行和第 4 行都是 false。

当程序执行第 16 行"object1=object2;"命令后,object1 的地址指向 object2 的地址,两者指向的是同一个存储地址,或者说是同一对象的不同变量名称。因此输出结果的第 7~8 行都返回 true。

程序第 22~23 行 str1 和 str2 通过 new 关键词调用 String()构造方法实现字符串的初始化,str1 和 str2 分配不同的存储地址,但值是相同的,对应输出结果第 12~13 行。在实际编程中会遇到大量的字符串比较,大多数情况是比较字符串的值,此时使用 equals()方法才能得到预期的正确结果。

"String str3＝"a";"是将字符串"a"转换为 String 类型并将存储地址赋予 String 包装类的实例 str3。按同样的方式将"a"转 String 类型后的同一地址赋给 str4,所以 str3 和 str4 的存储地址是相同的,对应输出结果第 17~18 行,返回都为 true。

包装类 Integer 赋值方式有如下三种。

(1) 采用"new Integer(int i)"命令,在新版本的 JDK 中已不推荐使用。

(2) 采用"Integer.valueOf(int i)"命令。

(3) 更普遍的使用方式是直接给 Integer 变量赋一个 int 型数值,int 型数值会隐式转换为 Integer 类型。

后面两种赋值方式是等效的。在输出结果的第 32 行 x==y 居然是 true。这是因为 Integer 中有一个静态内部类 IntegerCache,在类加载时,它会把[-128,127]的值缓存起来,而 x=Integer.valueOf(1)或 Integer a=1 这样的赋值方式会首先调用 Integer 类中的静态 valueOf()方法,这个方法会尝试从缓存中取值,如果在这个范围之内就不用重新构造一个对象并分配地址。如果多次调用,会取得同一个对象的引用,所以恒等比较返回 true。这同样也就能解释输出结果的第 46 和 48 行为什么赋值 100 的恒等比较返回 true,赋值 200 的恒等比较返回 false。

2.8 方法重载

方法重写和方法重载是两个完全不同的概念。方法重写是指子类覆盖父类的同名方法,属于在运行时进行绑定,属于动态多态。方法重载是指同一个类中有多个同名方法,是在编译时绑定的,属于静态多态。

【注1】 方法重载必须同时满足以下条件。
(1) 同一个类中至少有两个方法的名称相同。
(2) 同名方法的参数数量或参数类型不同。
【注2】 方法重载与返回类型无关。

【FirstActivity.java】
```
01    public class FirstActivity extends Activity
02    {
03        @Override
04        public void onCreate(Bundle savedInstanceState)
05        {
06            super.onCreate(savedInstanceState);
07            setContentView(R.layout.main);
08
09            Test test1 = new Test();
10            test1.test(5);
11            test1.test(5l);
12            test1.test(5.0);
13            test1.test(5.0, 6.0);
14        }
15    }
16
17    class A
18    {
19        void test(int x)
20        {
21            Log.i("xj", "test(int):" + x);
22        }
23
24        void test(long x)
```

```
25          {
26              Log.i("xj", "test(long):" + x);
27
28          }
29
30          void test(double x)
31          {
32              Log.i("xj", "test(double):" + x);
33          }
34
35          String test(double x, double y)
36          {
37              Log.i("xj", "test(double):x = " + x + ", y = " + y);
38              return "test(double):x = " + x + ", y = " + y;
39          }
40      }
```

第17～40行定义了类Test,其中定义了4个test()方法,它们都属于test()方法重载。前3个test()方法的参数类型不一样,第4个test()方法的参数数量和前三个不一样。如果方法名、参数类型和参数数量都相同,只是返回数据类型不同,不属于方法重载,Android Studio会用红色波浪线将相应代码标识为错误。

第9～13行创建类实例test1,并依次调用test()方法,传递的实参分别为整型、长整型和双精度数据类型。程序运行结果如下:

```
I: test(int):5
I: test(long):5
I: test(double):5.0
I: test(double):x = 5.0, y = 6.0
```

从结果可以看出,test()方法传递的实参数据类型和数量与Test类中test()重载方法的形参数据类型和数量一一对应,并执行相应重载方法中的代码。

将第19～22行代码加注释,即取消只有一个int型形参的test()方法,重新运行程序,运行结果如下:

```
I: test(long):5
I: test(long):5
I: test(double):5.0
I: test(double):x = 5.0, y = 6.0
```

当运行到第10行,传递的实参为整型值5,此时Test类中没有形参为整型的test()方法,Java会按照自动类型转换方式选择调用形参为长整型的test方法。

【注】 自动类型转换也称隐式类型转换,其转换顺序为:
byte→short→char→int→long→float→double。

2.9 代 码 块

在实际编写程序过程中,初学者往往不清楚代码的放置位置和执行顺序,其中一个原因是没有理解代码块的作用。

【FirstActivity.java】
```
01  public class FirstActivity extends Activity
02  {
03      int i;
04      //     i = 0;
05      int j = 0;
06
07      {
08          int m;
09          m = 0;
10      }
11
12      @Override
13      public void onCreate(Bundle savedInstanceState)
14      {
15          super.onCreate(savedInstanceState);
16          setContentView(R.layout.main);
17
18          Log.i("xj", "onCreate方法中");
19          Test test1 = new Test("张三");
20          Test test2 = new Test("李四");
21      }
22
23      {
24          Log.i("xj", "本代码块放置在onCreate()方法之后");
25      }
26
27      static
28      {
29          Log.i("xj", "FirstActivity中的静态代码块");
30      }
31  }
32
33  class Test
34  {
35      Test(String name)
36      {
37          Log.i("xj", "Test中的构造方法, name = " + name);
38      }
39
40      String str1 = "字符串";
41      static String str2 = "static字符串";
```

```
42
43    {
44        Log.i("xj", "Test 中的代码块:" + str1);
45    }
46
47    static
48    {
49        Log.i("xj", "Test 中的静态代码块:" + str2);
50        //Log.i("xj", "Test 中的静态代码块" + str1);
51    }
52 }
```

程序运行结果如下：

```
I: FirstActivity 中的静态代码块
I: 本代码块放置在 onCreate()方法之后
I: onCreate()方法中
I: Test 中的静态代码块:static 字符串
I: Test 中的代码块:字符串
I: Test 中的构造方法, name = 张三
I: Test 中的代码块:字符串
I: Test 中的构造方法, name = 李四
```

程序第 3 行是声明变量 i，第 5 行是声明变量 j 并且同时赋值 0。这两行都没有错误，也能正常执行。但第 4 行如果去掉注释就会提示"错误：需要<标识符>"。第 8～9 行却能正常执行，与第 3～4 行的区别在于多了一对花括号"{ }"。使用花括号括起的代码叫代码块。

【注】 定义的类中可以声明成员变量（并同时赋值）、定义方法，但不能单独给成员变量赋值或调用方法，除非放置在代码块中。

第 13 行的 onCreate()方法可视为定义方法，其方法体为代码块。同样第 24 行放置在代码块中，所以也能正常执行。

第 23～25 行的代码块也可以视为方法名称为匿名的方法。代码块会在类中定义的方法（含构造方法）之前执行。

第 27～30 行用 static 修饰的代码块也称静态代码块。静态代码块会随着类的加载而运行，而且只执行一次。所以第 29 行和第 49 行会优先执行，且只执行一次。

第 49 行能正常执行，但第 50 行却不行，原因是变量 str2 是静态变量而变量 str1 不是静态变量。在静态代码块中不能调用非静态变量或对象，这也同样适用于静态方法。

【注】 代码块的执行顺序：静态代码块→代码块→构造方法→普通方法。

第 3 章　Android 常用布局

目前使用 Android Studio 开发设计 UI(User Interface,用户接口)时还无法提供类似 Visual Studio 所见即所得的图形界面设计方式,但依靠线性布局(LinearLayout)、表格布局(TableLayout)、相对布局(RelativeLayout)、帧布局(FrameLayout)、绝对布局(AbsoluteLayout)、网格布局(GridLayout)和约束布局(ConstraintLayout)等已经能开发出各式各样 UI 界面。本章节通过案例来学习各种布局的特点和相关属性设置。

在 Android 的 UI 开发中需要了解长度单位的几种表示方式。

视频讲解

3.1　Android 长度单位

Android 布局设计的长度单位没有完全统一。常见的单位有 px、dp、sp、pt、mm 和 in 共 6 种。在布局文件的长度相关属性值中输入数字后,弹出智能提示中的长度单位,如图 3-1 所示。智能弹出提示框中会显示 6 种长度单位供开发人员选择。

以下是与长度相关的技术术语。

(1) px:即像素(pixels),1px 代表屏幕上一个物理的像素点。

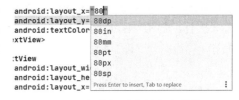

图 3-1　智能提示中的长度单位

(2) dp:独立像素密度(Density Independent Pixels),早期叫 dip),与像素无关。

(3) sp:主要用于设置字体尺寸,会随着系统的字体大小而改变,即同样大小的 dp 和 sp 字体,在 Android 设置中改变字体大小后,以 sp 为单位的字体会随系统字体大小改变而改变,以 dp 为单位的字体大小不会改变。正常字体 1dp=1sp,大字体和超大字体 1sp>1dp。以下是布局文件代码。

【main.xml】
```
01  <?xml version = "1.0" encoding = "UTF - 8"?>
02  < LinearLayout xmlns:android = "http://schemas.android.com/apk/res/android"
03      android:layout_width = "match_parent"
04      android:layout_height = "match_parent"
05      android:orientation = "vertical">
06
07      < TextView
08          android:layout_width = "wrap_content"
```

```
09          android:layout_height = "wrap_content"
10          android:text = "Hello World! 你好,安卓!18sp"
11          android:textSize = "18sp" />
12
13      < TextView
14          android:layout_width = "wrap_content"
15          android:layout_height = "wrap_content"
16          android:text = "Hello World! 你好,安卓!18dp"
17          android:textSize = "18dp" />
18  </LinearLayout >
```

以上代码按系统默认字体大小的效果如图 3-2 所示。

系统字体改成大字体后的效果如图 3-3 所示。

（4）in：英寸，1in＝2.54cm，一般用于屏幕对角线尺寸单位。

（5）pt：磅，1in/72 的长度，1pt＝1in * 2.54cm/72in≈0.035cm。

（6）分辨率：如果屏幕的分辨率是 1080 * 1920，是指水平方向上的像素数是 1080px，垂直方向上像素数是 1920px，屏幕分辨率如图 3-4 所示，根据勾股定律对角线则为 2203px。

图 3-2　系统默认字体大小的效果

图 3-3　系统字体改成大字体后的效果

图 3-4　屏幕分辨率

（7）屏幕像素密度：图 3-4 的对角线的像素数为 2203px，如果是 5 英寸屏（指对角线尺寸），屏幕像素密度为 2203÷5＝440；如果是 6 寸屏，屏幕像素密度为 2203÷6＝367。如此一来就会有很多不同的屏幕像素密度，同样的图片在屏幕中显示所占比例也就不同。为此 Android 引入像素密度与逻辑密度的概念。

（8）像素密度与逻辑密度：像素密度（dot per inch，dpi）就是每英寸的像素点数，不同的像素密度对应不同的 Android dpi 名称。如像素密度是 160，意思是每英寸像素数 160px，对应的 dpi 名称为 mdpi。Android Studio 在构建项目时会自动建立一个名为 HelloAndroid\app\src\main\res\mipmap-mdpi 的目录，目录中默认提供的图片分辨率为 48×48。Android Studio 同时也会建立其他 Android dpi 名称的目录，让不同分辨率、不同尺寸的 Android 设备自行调用不同目录中的图片文件，以保证不同参数的屏幕尽可能显示相似界面。像素密度是 40 的倍数。像素密度与逻辑密度如表 3-1 所示。

表 3-1　像素密度与逻辑密度

Android dpi 名称	分辨率	默认图片尺寸	像素密度	比例	逻辑密度
xxxhdpi	3840 * 2160	192 * 192	640	16	4
xxhdpi	1920 * 1080	144 * 144	480	12	3
xhdpi	1280 * 720	96 * 96	320	8	2
hdpi	480 * 800	72 * 72	240	6	1.5
mdpi	480 * 320	48 * 48	160	4	1
ldpi	320 * 240	36 * 36	120	3	0.75

逻辑密度是以 160dpi 为基准，其他像素密度与 160dpi 的比值，或者是像素密度对应的比例值除以 4，也是 dp 转 px 的系数。

dp 与 px 的换算关系为：**px＝dp * dpi/160**，如 160dpi 的 mdpi，1dp * 160dpi/160＝1px，对于基准 160dpi 而言，逻辑密度 density＝1(160dpi/160dpi 或者比例 4/4)。又如 240dpi 的 hdpi，1dp * 240dpi/160＝1.5px，对于基准 160dpi 而言，逻辑密度 density＝1.5(240dpi/160dpi 或者比例为 6/4)。换言之，要将 mdpi 的图片在 hdpi 上也能同比例显示，只需将 mdpi 下的图片放大 1.5 倍即可。在实际的设计中还可以使用 Java 代码获取屏幕分辨率然后换算比例进行屏幕动态布局，以此保证在不同分辨率的屏幕下都能按同样比例显示。以下是获取屏幕相关尺寸参数代码。

```java
【FirstActivity.java】
01  public class FirstActivity extends Activity
02  {
03
04      @Override
05      public void onCreate(Bundle savedInstanceState)
06      {
07          super.onCreate(savedInstanceState);
08          setContentView(R.layout.main);
09
10          String str = "";
11          DisplayMetrics dm = new DisplayMetrics();
12          dm = this.getApplicationContext().getResources().getDisplayMetrics();
13          str += "屏幕分辨率为:" + dm.widthPixels
                 + " * " + dm.heightPixels + "\n";
14          str += "水平方向分辨率:" + dm.widthPixels + "px\n";
15          str += "垂直方向分辨率:" + dm.heightPixels + "px\n";
16          str += "逻辑密度:" + dm.density + "\n";
17          str += "xdpi:" + dm.xdpi + "像素/英寸\n";
18          str += "ydpi:" + dm.ydpi + "像素/英寸\n";
19          Log.i("xj", str);
20      }
21  }
```

程序运行结果如下：

```
I:屏幕分辨率为:1080 * 1776
  水平方向分辨率:1080px
  垂直方向分辨率:1776px
  逻辑密度:3.0
  xdpi:480.0 像素/英寸
  ydpi:480.0 像素/英寸
```

将上面运行结果 xdpi 的 480 除以基准像素密度 160,得到的结果 3 就是逻辑密度。

3.2 线性布局

视频讲解

线性布局是 Android 早期开发版本的默认布局,使用 LinearLayout 标签,通过设置 android:orientation 属性值为 horizontal(水平)或 vertical(垂直)来将其内的控件按照水平方向或垂直方向依次排列。排列的控件不用指定位置的相关属性(简单即是美的体现),控件显示位置与在线性布局中出现的先后顺序相关。线性布局可以互相嵌套形成更复杂的结构。布局文件代码如下:

【main.xml】
```
01  <?xml version = "1.0" encoding = "UTF - 8"?>
02  < LinearLayout xmlns:android = "http://schemas.android.com/apk/res/android"
03      xmlns:tools = "http://schemas.android.com/tools"
04      android:layout_width = "match_parent"
05      android:layout_height = "match_parent"
06      android:orientation = "vertical">
07
08      < LinearLayout
09          android:layout_width = "match_parent"
10          android:layout_height = "wrap_content"
11          android:orientation = "horizontal">
12
13          < TextView
14              android:layout_width = "wrap_content"
15              android:layout_height = "wrap_content"
16              android:background = "@android:color/holo_red_dark"
17              android:text = "第一列红色"
18              android:textColor = "# ffffff"
19              android:textSize = "25sp" />
20
21          < TextView
22              android:layout_width = "wrap_content"
23              android:layout_height = "wrap_content"
24              android:background = "# 00aa00"
25              android:text = "绿色"
26              android:textColor = "# ffffff"
27              android:textSize = "25sp" />
28
```

```
29      <TextView
30          android:layout_width = "200dp"
31          android:layout_height = "wrap_content"
32          android:background = "#0000aa"
33          android:text = "第三列蓝色"
34          android:textColor = "#ffffff"
35          android:textSize = "25sp" />
36  </LinearLayout>
37
38  <LinearLayout
39      android:layout_width = "match_parent"
40      android:layout_height = "match_parent"
41      android:orientation = "vertical">
42
43      <TextView
44          android:layout_width = "wrap_content"
45          android:layout_height = "wrap_content"
46          android:background = "#0088ee"
47          android:text = "第一行为wrap_content"
48          android:textColor = "#ffffff"
49          android:textSize = "25sp" />
50
51      <TextView
52          android:layout_width = "match_parent"
53          android:layout_height = "wrap_content"
54          android:background = "#191970"
55          android:text = "第二行为match_parent"
56          android:textColor = "#ffffff"
57          android:textSize = "25sp" />
58
59      <TextView
60          android:layout_width = "wrap_content"
61          android:layout_height = "wrap_content"
62          android:background = "#11ff33"
63          android:text = "第三行wrap_content\n3333333333333333333333333333333333"
64          android:textColor = "#ffffff"
65          android:textSize = "25sp" />
66
67      <TextView
68          android:layout_width = "match_parent"
69          android:layout_height = "wrap_content"
70          android:background = "#191970"
71          android:text = "第四行match_parent\n4444444444444444444444444444444444"
72          android:textColor = "#ffffff"
73          android:textSize = "25sp" />
74
75  </LinearLayout>
76
77 </LinearLayout>
```

线性布局的运行结果如图 3-5 所示。

从布局代码和组件树(见图 3-6)所示的结构可以看出,整个布局的结构为,最外层是一个垂直线性布局,内嵌一个水平线性布局(含 3 个 TextView)和一个垂直线性布局(含 4 个 TextView)。TextView 控件的 android:layout_width 属性常用值有 match_parent(早期版本为 fill_parent,将 TextView 控件宽度设置为父容器宽度)、wrap_content(按 TextView 的显示文本内容长度来设置宽度,如果文本内容显示宽度超过父容器宽度则将宽度设为父容器宽度,多余的文本在下一行中显示)和具体数值宽度(如第 30 行的 200dp)。android:layout_height 属性用于设置高度,与设置宽度概念类似。

图 3-5 线性布局的运行结果

图 3-6 组件树

显示界面标题栏下的第一行分别显示红、绿、蓝 3 种颜色的 TextView,前两个 TextView 按文字长度显示,第三个 TextView 定义的宽度为 200dp,大于文字长度,所以 TextView 控件中文字离控件右边缘有一段距离。

后面几行显示了不同文字长度的 TextView 在 android:layout_width 宽度属性分别设置为 match_parent 和 wrap_content 时的显示差异。读者可修改代码观察运行结果中 TextView 的宽度和折行变化。

在图 3-6 的组件树右侧单击 ◉ 图标按钮可以修改控件是否可见。⚠ 图标提示控件代码有警示信息。本案例中将 TextView 中的文字直接以字符串形式赋予了 android:text,而 Android Studio 建议采用"@string/字符串变量名"的方式定义 TextView 的值。在 Code 视图中警示方式会变为将相应属性背景色变为黄色,如图 3-7 所示。在 Design 视图中单击橙色三角形 ⚠,在弹出信息栏中单击 Fix 按钮(如果是 Code 视图,则将光标置于字符串中,按 lt+Enter 快捷键,在上下文菜单中选择 Extract string resource 菜单),在弹出的界面中输入字符串变量名,系统自动在 strings.xml 中注册相应的字符串变量名和对应的字符串值。如果弹出的三角形是红色,代表控件的属性设置中有错误。开发人员可以用鼠标拖动组件树中的控件位置来改变控件显示顺序,Code 视图中的代码也会自动调整顺序。右击控件树中的线性布局,在弹出的快捷菜单中可选择转换为其他布局方式。

Code 视图如图 3-7 所示,android:background 属性用来设置背景色。android:textColor 属

性用来设置文字颜色,颜色值使用 6 位十六进制数表示,每 2 位为一组,分别表示红、绿、蓝,合成的颜色效果在左侧行号后显示为颜色方块或在 Design 视图中查看效果。

用鼠标双击 Code 视图行号后的颜色方块弹出调色盘,如图 3-8 所示,可实现可视化的颜色调配选择。如果在图 3-8 中选择 Resources 选项卡,可选择系统自定义的颜色,如第 16 行的"@android:color/holo_red_dark",代表使用 Android 自定义的颜色。Design 视图中显示的效果与实际运行结果可能会有差异,以实际运行结果为准。

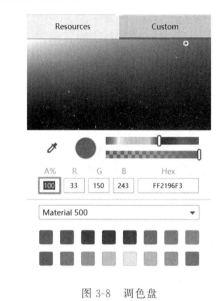

图 3-7　Code 视图　　　　　　　　　　图 3-8　调色盘

3.3　边 线 和 角

所有的布局方式(含 LinearLayout)设定的边界都是没有线条标识的,如果想给相关布局画出边线可以采用以下方式:

```
【main.xml】
01  <?xml version = "1.0" encoding = "UTF-8"?>
02  < LinearLayout xmlns:android = "http://schemas.android.com/apk/res/android"
03      android:layout_width = "match_parent"
04      android:layout_height = "match_parent"
05      android:background = "@drawable/shape_conner"
06      android:gravity = "center"
07      android:orientation = "vertical">
08
09      < TextView
10          android:layout_width = "wrap_content"
11          android:layout_height = "wrap_content"
12          android:background = "#ffffff"
13          android:text = "@string/hello"
14          android:textSize = "15dp" />
15  </LinearLayout >
```

在线性布局标签属性中添加第 5 行的代码，设置线性布局背景使用 drawable 目录下的 shape_conner.xml 文件（此时 shape_conner.xml 文件被当成一个图片文件使用）。

```xml
【shape_conner.xml】
01  <?xml version = "1.0" encoding = "UTF - 8"?>
02  < shape xmlns:android = "http://schemas.android.com/apk/res/android">
03      <!-- 内部背景色 -->
04      < solid android:color = "#5f5fdc"/>
05      <!-- 边角半径 -->
06      < corners
07          android:bottomLeftRadius = "30dp"
08          android:bottomRightRadius = "30dp"
09          android:topLeftRadius = "10dp"
10          android:topRightRadius = "10dp"/>
11      <!-- 边线颜色和边线宽度 -->
12      < stroke
13          android:width = "5dp"
14          android:color = "#ff0000"/>
15  </shape>
```

第 4 行定义了背景色。

第 6~10 行定义了屏幕 4 个角的转角半径，如果半径设为 0 则为直角。

第 13 行定义了线条的宽度。

第 14 行定义了线条的颜色。

给布局添加边线和角的运行结果如图 3-9 所示。

图 3-9　给布局添加边线和角的运行结果

当前屏幕为圆角的手机越来越多，而大多数模拟器的 4 个角是直角，如果设计时就处理边角显示适配，则可以考虑采用此方法。

3.4 layout_weight

Android 布局中设置控件的宽度一般使用 android:layout_width 属性,有时会采用属性 android:layout_weight 与 android:layout_width 配合使用。android:layout_weight 的作用是设定同一父容器内控件的长度或宽度的占比。先通过一个布局代码来看实际运行结果,再分析与显示结果不同的原因。

【main.xml】
```
01  <?xml version = "1.0" encoding = "UTF-8"?>
02  <LinearLayout xmlns:android = "http://schemas.android.com/apk/res/android"
03      android:layout_width = "match_parent"
04      android:layout_height = "match_parent"
05      android:orientation = "vertical">
06
07      <LinearLayout
08          android:layout_width = "match_parent"
09          android:layout_height = "wrap_content"
10          android:orientation = "horizontal">
11
12          <Button
13              android:layout_width = "wrap_content"
14              android:layout_height = "wrap_content"
15              android:layout_weight = "1"
16              android:text = "按钮 1"
17              android:textSize = "20sp" />
18
19          <Button
20              android:layout_width = "wrap_content"
21              android:layout_height = "wrap_content"
22              android:layout_weight = "2"
23              android:text = "按钮 2"
24              android:textSize = "20sp" />
25      </LinearLayout>
26
27      <LinearLayout
28          android:layout_width = "match_parent"
29          android:layout_height = "wrap_content"
30          android:orientation = "horizontal">
31
32          <Button
33              android:layout_width = "match_parent"
34              android:layout_height = "wrap_content"
35              android:layout_weight = "1"
36              android:text = "按钮 3"
37              android:textSize = "20sp" />
38
39          <Button
```

```
40              android:layout_width = "match_parent"
41              android:layout_height = "wrap_content"
42              android:layout_weight = "2"
43              android:text = "按钮 4"
44              android:textSize = "20sp" />
45      </LinearLayout>
46
47  </LinearLayout>
```

整个布局的结构是一个垂直线性布局内嵌两个水平线性布局,每个水平布局中又放置两个按钮,"按钮 1"和"按钮 3"的 layout_weight 设为 1,"按钮 2"和"按钮 4"的 layout_weight 设为 2。"按钮 1"宽度占屏幕的 1/3,"按钮 2"宽度占屏幕的 2/3,这与设想的相符。"按钮 3"宽度占屏幕的 2/3,"按钮 4"宽度占屏幕的 1/3,这是怎么回事呢?产生差异的原因是按钮布局 android:layout_width 属性是 wrap_content 还是 match_parent。

(1) 当 android:layout_width = "wrap_content"时(假设按钮的文本内容长度没有超过屏幕占比),两个按钮占屏幕一行,每个按钮按各自占比设置宽度,如此例中 layout_weight 分别为 1 和 2,则总和为 3,"按钮 1"占 1/3,"按钮 2"占 2/3。

(2) 当 android:layout_width = "match_parent"时,各按钮的宽度等于父容器宽度加上剩余空间的占比。设父容器宽度为 L,"按钮 3"和"按钮 4"的 android:layout_width = "match_parent",所以两个按钮宽度都应该为 L,剩余宽度就为父容器宽度减去两个按钮的宽度:L−(L+L)=−L。"按钮 1"占 1/3,所以"按钮 3"的实际宽度是 L(父容器宽度)+(−L)(剩余宽度)*1/3=L+(−L)*1/3=2L/3。同理,"按钮 4"的实际宽度为 L/3。

由此可以看出,Android 在长度设置上除了长度单位不同外,还要考虑不同属性之间的影响。layout_weight 属性并不能精确地控制控件的宽度(或高度),还会受控件内文字长度的影响(即使文字长度未超过屏幕占比)。如果想精确控制各控件的长度对齐,需考虑使用其他布局。

layout_width 结合 layout_weight 运行结果如图 3-10 所示。

图 3-10 layout_width 结合 layout_weight 运行结果

【提问】 删除第 35 和 42 行将如何显示?

3.5 绝对布局

绝对布局使用 android:layout_x 和 android:layout_y 来设定屏幕水平方向和垂直方向坐标,这种定位方式简单直接,但对于不同分辨率的屏幕,绝对布局的显示效果会有差异,这也是不推荐使用绝对布局的原因。一种解决方式是先获取屏幕的分辨率,然后按照百分比计算绝对布局的 x、y 坐标。在 Android Studio 中查看绝对布局源码,单词 AbsoluteLayout

会出现一条中画线,将鼠标指针放在单词 AbsoluteLayout 上,弹出弃用提示,如图 3-11 所示。按下快捷键 Alt+Shift+Enter,AbsoluteLayout 标签属性中会自动添加一行属性 tools:ignore="Deprecated"来忽略弃用提示,此时会看到单词 AbsoluteLayout 上的中画线消失。"tools:"标识并不影响布局设计,只是对界面设计人员起到辅助的作用。

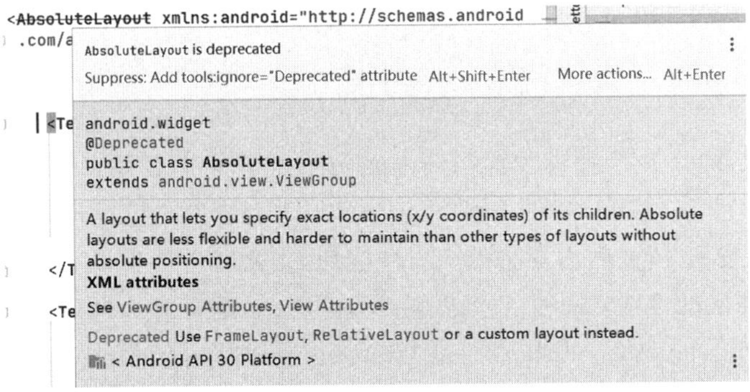

图 3-11　弃用提示

以下是绝对布局的源码和运行结果。

```
【main.xml】
01  <?xml version = "1.0" encoding = "UTF-8"?>
02  <AbsoluteLayout xmlns:android = "http://schemas.android.com/apk/res/android"
03      xmlns:tools = "http://schemas.android.com/tools"
04      android:layout_width = "match_parent"
05      android:layout_height = "match_parent"
06      tools:ignore = "Deprecated">
07
08      <TextView
09          android:layout_width = "wrap_content"
10          android:layout_height = "wrap_content"
11          android:layout_x = "0px"
12          android:layout_y = "0px"
13          android:text = "@string/hello">
14      </TextView>
15
16      <TextView
17          android:layout_width = "wrap_content"
18          android:layout_height = "wrap_content"
19          android:layout_x = "80px"
20          android:layout_y = "80px"
21          android:text = "@string/action">
22      </TextView>
23
24      <TextView
25          android:layout_width = "wrap_content"
26          android:layout_height = "wrap_content"
27          android:layout_x = "150px"
28          android:layout_y = "150px"
```

```
29         android:text = "@string/hello">
30     </TextView>
31
32     <TextView
33         android:layout_width = "wrap_content"
34         android:layout_height = "wrap_content"
35         android:layout_x = "140px"
36         android:layout_y = "145px"
37         android:text = "@string/collision"
38         android:textColor = "#ff00ff">
39     </TextView>
40
41     <TextView
42         android:layout_width = "wrap_content"
43         android:layout_height = "wrap_content"
44         android:layout_x = "0px"
45         android:layout_y = "750px"
46         android:text = "@string/lastLine"
47         android:textColor = "#000000">
48     </TextView>
49
50 </AbsoluteLayout>
```

第11~12行TextView的定位为(0,0),从运行结果上可以看出原点在标题栏下方的显示区左上角(不是整个屏幕的左上角)。

第24~39行中定义的两个TextView坐标非常接近,显示的文本也就有部分重叠。这是其他布局方式难以达到的效果(帧布局和约束布局除外)。这也是为什么还是有一部分开发人员喜欢使用绝对布局,特别是针对单一设备,此时不用考虑屏幕尺寸和分辨率带来的差异。

绝对布局运行结果如图3-12所示。

图3-12 绝对布局运行结果

3.6 相对布局

在新版的 Android Studio 中，相对布局已经归入 Legacy 控件栏中，意味着以后可能会弃用。这里仍然讲解相对布局，其一是因为很多网络上的案例仍在使用相对布局；其二是相对布局中的一些属性和概念也可以用于其他布局中。相对布局的常用属性有以下 4 类。

（1）当前控件与父容器的相对位置，属性值为 true 或者 false。与父容器相对位置如图 3-13 所示，粗线代表控件对齐的边。layout_alignParentStart 属性默认对应 layout_alignParentLeft，layout_alignParentEnd 属性默认对应 layout_alignParentRight，此时默认布局方向是 left-to-right。如果采用 right-to-left 布局方向，则 layout_alignParentStart 属性对应 layout_alignParentRight，layout_alignParentEnd 属性对应 layout_alignParentLeft。本书后续涉及带 Start 和 End 的属性按默认 left-to-right 布局方向解释为 Left 和 Right。新版 Android Studio 推荐使用 Start 和 End 替代 Left 和 Right。

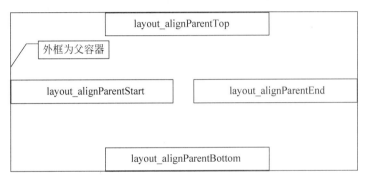

图 3-13　与父容器相对位置

（2）当前控件与参考控件的相对位置，属性值必须为参考控件 id 的引用名"@id/id-name"。与参考控件相对位置如图 3-14 所示。

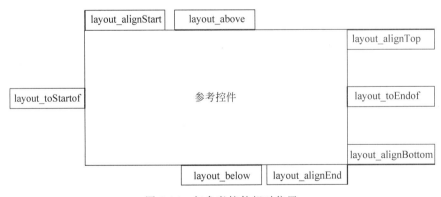

图 3-14　与参考控件相对位置

（3）属性值是具体的长度或像素，如 android:layout_marginTop="100dp"，指明当前控件离父容器上边缘 100dp 距离。

（4）定义控件边界的空白宽度和控件内部填充宽度属性，相关属性如图 3-15 所示，属性值为长度或像素值。

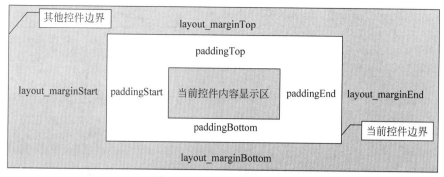

图 3-15 margin 与 padding

```
【main.xml】
01  <?xml version = "1.0" encoding = "UTF-8"?>
02  < RelativeLayout xmlns:android = "http://schemas.android.com/apk/res/android"
03      android:layout_width = "match_parent"
04      android:layout_height = "match_parent">
05
06      < TextView
07          android:id = "@ + id/textView"
08          android:layout_width = "match_parent"
09          android:layout_height = "wrap_content"
10          android:layout_marginTop = "100dp"
11          android:text = "@string/layout" />
12
13      < EditText
14          android:id = "@ + id/editText1"
15          android:layout_width = "match_parent"
16          android:layout_height = "wrap_content"
17          android:layout_below = "@ + id/textView"
18          android:layout_alignParentRight = "true" />
19
20      < Button
21          android:id = "@ + id/button1"
22          android:layout_width = "wrap_content"
23          android:layout_height = "wrap_content"
24          android:layout_below = "@ + id/editText1"
25          android:layout_alignParentEnd = "true"
26          android:layout_marginEnd = "80dp"
27          android:text = "@string/button1" />
28
29      < Button
30          android:id = "@ + id/button2"
31          android:layout_width = "wrap_content"
32          android:layout_height = "wrap_content"
33          android:layout_alignBottom = "@ + id/button1"
34          android:layout_alignParentStart = "true"
35          android:text = "@string/button2" />
36
```

```
37          <Button
38              android:id = "@ + id/button3"
39              android:layout_width = "wrap_content"
40              android:layout_height = "wrap_content"
41              android:layout_alignParentBottom = "true"
42              android:layout_toEndOf = "@ + id/button2"
43              android:text = "@string/button3" />
44
45    </RelativeLayout>
```

第 10 行指明 TextView 与父容器的上边缘间隔为 100dp。

第 24 行指明 button1 在文本框 editText1 的下方。第 25 行指明 button1 与父容器右对齐。第 26 行让 button1 右侧与对齐边线保持 80dp 的间隔。

第 33 行指明 button2 与 button1 底部对齐。第 34 行指明 button2 与父容器左对齐。

第 41 行指明 button3 对齐父容器的底部，第 42 行指明 button3 的左边与 button2 的右边缘对齐。相对布局运行结果如图 3-16 所示。

案例库中还列出了使用 Java 代码来实现动态设定相对布局，由于还未讲解按钮监听器的使用，读者可在学习相应章节后自行查看、使用相应代码。

【提问】 第 42 行换成 android:layout_toEndOf = "@ + id/button1"会如何？

图 3-16　相对布局运行结果

如果将第 42 行的 button2 改为 button1，运行后会发现 button3 不见了。这是因为 button1 在第 25 行指明是右对齐屏幕右边缘，虽然第 26 行留出了 80dp 的间隔，但其他控件如果与 button1 右边缘对齐，还是要以没有间隔的位置为准，所以 button3 的位置在屏幕右侧边缘之外导致看不见了。

视频讲解

3.7　帧　布　局

帧布局为每个加入其中的控件创建一个区域（称为一帧），这些帧会根据 layout_gravity 属性执行相应对齐。未设置 layout_gravity 属性值时，控件默认在父容器的左上角。符号"|"用于定义同时拥有多个属性值。layout_gravity 属性值定位示意图如图 3-17 所示。

默认或者 start|top	top|center_horizontal	top|end
center_vertical|start	center_vertical|center_horizontal	center_vertical|end
bottom|start	bottom|center_horizontal	bottom|end

外框为父容器

图 3-17　layout_gravity 属性值定位示意图

【main.xml】
```xml
01  <?xml version = "1.0" encoding = "UTF-8"?>
02  <FrameLayout xmlns:android = "http://schemas.android.com/apk/res/android"
03      xmlns:tools = "http://schemas.android.com/tools"
04      android:layout_width = "match_parent"
05      android:layout_height = "match_parent">
06
07      <TextView
08          android:layout_width = "match_parent"
09          android:layout_height = "wrap_content"
10          android:text = "1111"
11          android:textColor = "#ff00ff"
12          android:textSize = "50sp" />
13
14      <TextView
15          android:layout_width = "match_parent"
16          android:layout_height = "wrap_content"
17          android:text = "2222"
18          android:textColor = "@android:color/black"
19          android:textSize = "40sp" />
20
21      <TextView
22          android:layout_width = "wrap_content"
23          android:layout_height = "wrap_content"
24
25          android:layout_gravity = "center_vertical|center_horizontal"
26          android:background = "@android:color/holo_orange_light"
27          android:text = "3333"
28          android:textSize = "30sp"
29          tools:layout_width = "wrap_content" />
30
31      <TextView
32          android:layout_width = "match_parent"
33          android:layout_height = "wrap_content"
34          android:background = "#666666"
35          android:layout_gravity = "bottom"
36          android:text = "4444"
37          android:textColor = "#ffffff"
38          android:textSize = "20sp" />
39      <TextView
40          android:layout_width = "wrap_content"
41          android:layout_height = "wrap_content"
42          android:background = "#666666"
43          android:layout_gravity = "center|end"
44          android:text = "5555"
45          android:textColor = "#ffffff"
46          android:textSize = "20sp" />
47  </FrameLayout>
```

第 7～19 行的两个 TextView 没有定义 layout_gravity，将叠加显示在父容器（在此例中 TextView 的父容器为 FrameLayout）的左上角。叠加的顺序是后定义的控件显示在之前定义的控件之上。

第 25 行 TextView 同时指定为垂直方向正中和水平方向正中，显示的效果是父容器的中心位置。

第 35 行定义为 bottom，此时变更为 bottom|end 也是相同的效果，因为当前 TextView 的 layout_width 属性定义为 match_parent，意味着 TextView 控件宽度与父容器等宽，所以再加上右对齐属性还是显示为与父容器等宽的右对齐。

案例中用两种方式（Android 自带颜色和十六进制）定义颜色，6 位十六进制数折合二进制是 24 位，也就是平时所说的 24 位色。帧布局运行结果如图 3-18 所示。

图 3-18 帧布局运行结果

视频讲解

3.8 表格布局

表格布局是按照行列的表格方式排列布局，其中 TableRow 用于同一行内多个控件对象的排列，如果没有定义 TableRow，则一个控件对象占用表格的一行。为了控制表格的拉伸和收缩，可设置以下属性。

(1) android:collapseColumns：设置需要被隐藏列的序号，相应列不可见。

(2) android:shrinkColumns：设置允许收缩列的序号。

(3) android:stretchColumns：设置允许拉伸列的序号。

【注】 列的序号是从 0 开始的。

下面的案例设计了一个规整的表格界面，代码如下：

```
【main.xml】
01  <?xml version = "1.0" encoding = "UTF-8"?>
02  <TableLayout xmlns:android = "http://schemas.android.com/apk/res/android"
03      android:layout_width = "match_parent"
04      android:layout_height = "match_parent"
05      android:background = "#C9E1F4"
06      android:stretchColumns = "0,1,2">
07
08      <TextView
09          android:background = "#2241EC"
10          android:gravity = "center"
11          android:text = "学生信息表"
12          android:textColor = "#FFFFFF"
13          android:textSize = "30dip" />
14      <TextView
15          android:background = "#2196F3"
16          android:gravity = "center"
```

```xml
17              android:text = "男生信息"
18              android:textColor = "#FFFFFF"
19              android:textSize = "30dip" />
20
21      <TableRow>
22
23          <TextView
24              android:layout_column = "0"
25              android:layout_margin = "4dip"
26              android:background = "#F8F7EE"
27              android:gravity = "center"
28              android:text = "学号" />
29
30          <TextView
31              android:layout_column = "1"
32              android:layout_margin = "4dip"
33              android:background = "#F8F7EE"
34              android:gravity = "center"
35              android:text = "姓名" />
36
37          <TextView
38              android:layout_margin = "4dip"
39              android:background = "#F8F7EE"
40              android:gravity = "center"
41              android:text = "出生地" />
42      </TableRow>
43
44      <TableRow>
45
46          <TextView
47              android:layout_margin = "4dip"
48              android:background = "#F8F7EE"
49              android:gravity = "center"
50              android:text = " 2021001 " />
51
52          <TextView
53              android:layout_margin = "4dip"
54              android:background = "#F8F7EE"
55              android:gravity = "left"
56              android:text = "张三" />
57
58          <TextView
59              android:layout_margin = "4dip"
60              android:background = "#F8F7EE"
61              android:gravity = "right"
62              android:text = "云南昆明" />
63      </TableRow>
64
65      <TableRow>
66
```

```
67        <TextView
68            android:layout_margin = "4dip"
69            android:background = "#F8F7EE"
70            android:gravity = "center"
71            android:text = " 2021002 " />
72
73        <TextView
74            android:layout_margin = "4dip"
75            android:background = "#F8F7EE"
76            android:gravity = "left"
77            android:text = "李四" />
78
79        <TextView
80            android:layout_margin = "4dip"
81            android:background = "#F8F7EE"
82            android:gravity = "right"
83            android:text = "北京" />
84    </TableRow>
85
86    <TableRow>
87
88        <TextView
89            android:layout_margin = "4dip"
90            android:background = "#F8F7EE"
91            android:gravity = "center"
92            android:text = "2021003" />
93
94        <TextView
95            android:layout_margin = "4dip"
96            android:background = "#F8F7EE"
97            android:gravity = "left"
98            android:text = "王五" />
99
100        <TextView
101            android:layout_margin = "4dip"
102            android:background = "#F8F7EE"
103            android:gravity = "right"
104            android:text = "四川成都" />
105    </TableRow>
106 </TableLayout>
```

第 6 行指明表格布局的 0~2 列是可拉伸的，本案例中表格布局有 3 列，默认这 3 列平分父容器宽度。当某列有单元格的字符超出平分表格列格宽度时，此列的宽度会自动扩展，其他列相应收缩，直至其左侧列宽度等于文本宽度或者其右侧列被挤出父容器之外。如果删除此行，则后续 TableRow 中控件按指定宽度或默认 warp_conent 显示。

第 8~19 行定义的 TextView 没有在 TableRow 标签内。表格布局中没有在 TableRow 标签内的控件默认都独占一行，所以两个 TextView 分别占了两行。

第 21~42 行属于 TableRow 标签范围，TableRow 标签内定义的 TextView 都在同

一行。

第27行的android:gravity用于TextView控件内部的文字对齐，android:layout_gravity用于当前控件对父容器的对齐。详细讲解参见"4.1.3 layout_gravity与gravity"中的案例。

表格布局运行结果如图3-19所示。

图3-19　表格布局运行结果

下面的案例设计了一个不规则表格界面，代码如下：

```
【main.xml】
01  <LinearLayout xmlns:android = "http://schemas.android.com/apk/res/android"
02      android:layout_width = "match_parent"
03      android:layout_height = "match_parent"
04      android:orientation = "vertical">
05
06      <TableLayout
07          android:id = "@ + id/tablelayout01"
08          android:layout_width = "match_parent"
09          android:layout_height = "wrap_content"
10          android:shrinkColumns = "1"
11          android:stretchColumns = "2">
12
13          <Button
14              android:id = "@ + id/button01"
15              android:layout_width = "wrap_content"
16              android:layout_height = "wrap_content"
17              android:text = "独自一行,不在TableRow中" />
18
19          <TableRow>
20
21              <Button
22                  android:id = "@ + id/button02"
23                  android:layout_width = "wrap_content"
```

```xml
24              android:layout_height = "wrap_content"
25              android:text = "表 1" />
26
27          <Button
28              android:id = "@ + id/button03"
29              android:layout_width = "wrap_content"
30              android:layout_height = "wrap_content"
31              android:text = "允许被收缩允许被收缩允许被收缩" />
32
33          <Button
34              android:id = "@ + id/button04"
35              android:layout_width = "wrap_content"
36              android:layout_height = "wrap_content"
37              android:text = "允许被拉伸允许被拉伸" />
38      </TableRow>
39  </TableLayout>
40
41  <TableLayout
42      android:id = "@ + id/tablelayout02"
43      android:layout_width = "match_parent"
44      android:layout_height = "wrap_content"
45      android:collapseColumns = "1">
46
47      <TableRow>
48
49          <Button
50              android:id = "@ + id/button05"
51              android:layout_width = "wrap_content"
52              android:layout_height = "wrap_content"
53              android:text = "表 2" />
54
55          <Button
56              android:id = "@ + id/button06"
57              android:layout_width = "wrap_content"
58              android:layout_height = "wrap_content"
59              android:text = "被隐藏列" />
60
61          <Button
62              android:id = "@ + id/button07"
63              android:layout_width = "wrap_content"
64              android:layout_height = "wrap_content"
65              android:text = "我是第三列" />
66      </TableRow>
67  </TableLayout>
68
69  <TableLayout
70      android:id = "@ + id/tablelayout03"
71      android:layout_width = "match_parent"
72      android:layout_height = "wrap_content"
73      android:stretchColumns = "1">
```

```
74
75            < TableRow >
76
77                < Button
78                    android:id = "@ + id/button08"
79                    android:layout_width = "wrap_content"
80                    android:layout_height = "wrap_content"
81                    android:text = "表 3" />
82
83                < Button
84                    android:id = "@ + id/button09"
85                    android:layout_width = "wrap_content"
86                    android:layout_height = "wrap_content"
87                    android:text = "填满剩余空白" />
88            </TableRow>
89        </TableLayout>
90
91 </LinearLayout>
```

第 10 行指明第 2 列是可收缩的(从 0 开始计算)。

第 11 行指明第 3 列是可拉伸的。

第 33～37 行定义的按钮文字比较多,其在表格布局的第 3 列,按钮宽度根据文字长度而拉伸,相应定义为可收缩的第 2 列按钮的宽度被压缩,其超出按钮宽度的文字将折行显示。

自第 41 行起重新定义了一个表格布局,其中第 45 行定义第 2 列可折叠隐藏。第 55～59 行定义的按钮被隐藏,所以表 2 只显示了两列。

第 69 行定义新的表格布局,其中第 73 行定义为第 2 列可拉伸,在 TableRow 中定义了两个按钮,第 2 个按钮的宽度虽然设置为 wrap_content,但因为缺第 3 列导致可拉伸的第 2 列填满剩余表格行宽度。

不规则表格布局运行结果如图 3-20 所示。

图 3-20　不规则表格布局运行结果

3.9　网格布局

在新版的 Android Studio 中,网格布局已经归入 Legacy 控件栏。官方更推荐表格布局作为类似场景中的布局。网格布局的默认行列高度和宽度是统一的,可以通过调整布局容器大小,设置 android:layout_width 或行列的权重来改变。以下案例设计一个简单计算器界面,具体代码如下:

```
01 < GridLayout xmlns:android = "http://schemas.android.com/apk/res/android"
02     android:layout_width = "wrap_content"
```

```
03        android:layout_height = "wrap_content"
04        android:columnCount = "4"
05        android:orientation = "horizontal"
06        android:rowCount = "6">
07
08        <EditText
09            android:id = "@ + id/result"
10            android:layout_columnSpan = "4"
11            android:layout_gravity = "fill" />
12
13        <Button
14            android:id = "@ + id/one"
15            android:text = "1" />
16
17        <Button
18            android:id = "@ + id/two"
19            android:text = "2" />
20
21
22        <Button
23            android:id = "@ + id/three"
24            android:text = "3" />
25
26        <Button
27            android:id = "@ + id/devide"
28            android:text = "/" />
29
30        <Button
31            android:id = "@ + id/four"
32            android:text = "4" />
33
34        <Button
35            android:id = "@ + id/five"
36            android:text = "5" />
37
38        <Button
39            android:id = "@ + id/six"
40            android:text = "6" />
41
42        <Button
43            android:id = "@ + id/multiply"
44            android:text = " × " />
45
46        <Button
47            android:id = "@ + id/seven"
48            android:text = "7" />
49
50        <Button
51            android:id = "@ + id/eight"
```

```
52            android:text = "8" />
53
54        < Button
55            android:id = "@ + id/nine"
56            android:text = "9" />
57
58        < Button
59            android:id = "@ + id/minus"
60            android:text = " - " />
61
62        < Button
63            android:id = "@ + id/zero"
64            android:layout_columnSpan = "2"
65            android:layout_gravity = "fill"
66            android:text = "0" />
67
68        < Button
69            android:id = "@ + id/point"
70            android:text = "." />
71
72        < Button
73            android:id = "@ + id/plus"
74            android:layout_rowSpan = "2"
75            android:layout_gravity = "fill"
76            android:text = " + " />
77
78        < Button
79            android:id = "@ + id/equal"
80            android:layout_columnSpan = "3"
81            android:layout_gravity = "fill"
82            android:text = " = " />
83    </GridLayout >
```

第 4 行定义网格布局有 4 列，第 6 行定义网格布局有 6 行，最后形成一个 6 行 4 列的网格。

第 10 行定义 EditText 可以拉伸 4 列宽度，再结合第 11 行实现 EditText 占 4 列宽度。同样第 74～75 行指明加号按钮占两行。

网格布局运行结果如图 3-21 所示。

从本案例可以看出，网格布局更适用于较为规整的行列表格，网格布局中的控件按指定的行列顺序依次排列，但相应的灵活性也比表格布局低。网格布局还存在控件间隙不一致的缺陷。

图 3-21　网格布局运行结果

3.10 约束布局

3.10.1 约束布局基础

视频讲解

从 Android Studio 2.3 版本起,约束布局是 Android Studio 布局文件的默认布局。其他布局方式在实现复杂一些的布局设计时存在多种或多个布局嵌套的情况,设备调用这样的布局文件就需要花费更多的时间。约束布局在灵活性和可视化方面比其他布局方式更胜一筹(与号称宇宙第一 IDE 的 Visual Studio 的图形化界面设计相比还有差距)。为减少布局嵌套,使用约束布局的属性更接近于相对布局属性。因此,很多资料在讲解约束布局时采用与相对布局类似的方式,不可避免地就要讲解一堆约束布局的定位属性。约束布局与其他布局的最大区别在于支持图形化拖放操作。可以在布局文件的 Design 视图中采用鼠标拖放操作结合属性栏窗口设置完成约束布局的界面设计,大幅简化布局代码输入和控件间定位关系的人为判断。

【注】 约束布局可以实现图形化拖放设计,但不是真正的所见即所得。Design 视图中的效果与实际运行结果还是有所不同的,经常出现的问题是定位属性设置错误或缺项。

选择布局文件 main.xml,在 Design 视图的工具栏中拖放两个 Button 按钮到 Design 界面中,如图 3-22 所示。

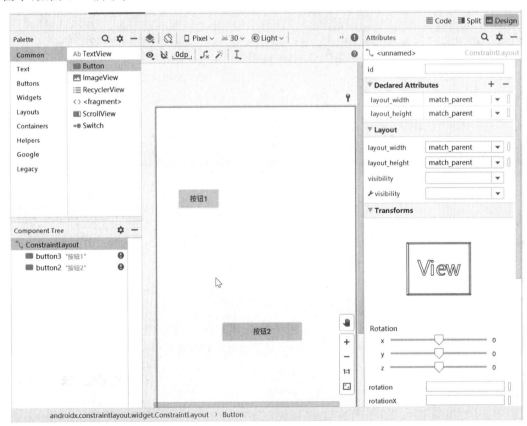

图 3-22 约束布局 Design 视图

此时由于未加入约束定位,组件树中的控件都会用红色惊叹号标识。运行程序,两个按钮会显示在屏幕左上角,坐标为(0,0)。

选中按钮控件,显示控件句柄,如图 3-23 所示。4 个角上的正方形句柄用于调整控件的尺寸,圆形句柄用于设置控件的定位。

图 3-23 控件句柄

在"按钮 1"左侧圆形句柄上按住鼠标左键并拖向屏幕左侧,此时"按钮 1"会自动靠到屏幕左侧,意味着"按钮 1"与屏幕左侧对齐。选中"按钮 1"再次向右拖动到如图 3-24 所示的位置,4 个方框圈出数字为 132,单位为 dp,代表"按钮 1"相对父容器(此时为屏幕左边缘)的距离为 132dp。也可以在属性栏或布局栏(图中右侧中间的 Constraint Widget)中修改数字改变距离值。以上操作对应两个属性:

```
app:layout_constraintStart_toStartOf = "parent"
android:layout_marginStart = "132dp"
```

注意,一个前缀是"app:",另一个前缀是"android:"。前者引用 app 目录下 build.gradle 文件中 androidx.constraintlayout:constraintlayout 定义的属性;后者引用系统定义的属性。

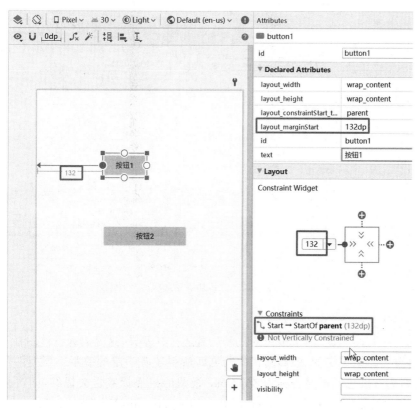

图 3-24 "按钮 1"相对父容器水平方向定位

同样,选择"按钮 1"上方的圆形句柄定位屏幕上边缘距离。重新运行程序,"按钮 1"将出现在设定的位置。"按钮 2"还是在屏幕左上角。可以采用同样的方式对"按钮 2"进行操作,也可以采用相对"按钮 1"的位置进行定位。例如,将"按钮 2"定位在"按钮 1"下方 50dp

位置，"按钮 2"右侧离屏幕右侧 100dp。为实现上述定位要求，先选中"按钮 2"上方的圆形句柄并拖动到"按钮 1"（此时"按钮 1"上会出现上、下两个圆形句柄）下方的圆形句柄，调整距离值为 50dp。如果进行此操作，鼠标左键释放时指向"按钮 1"区域而非圆形句柄，会弹出如图 3-25 的上下文菜单，可选择"按钮 2"是对齐"按钮 1"顶部还是底部。

选择"按钮 2"右侧的圆形句柄并拖动到屏幕右侧，调整距离值为 100dp，"按钮 2"的定位如图 3-26 所示。

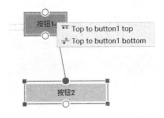

图 3-25　设定"按钮 2"相对"按钮 1"的垂直方向定位

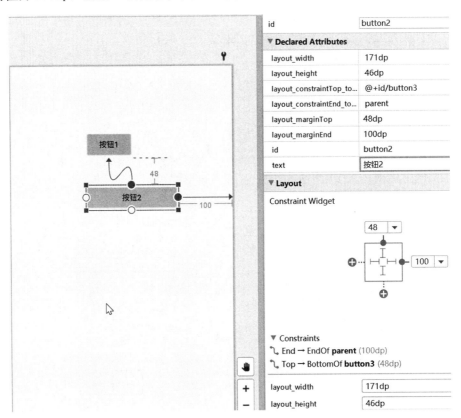

图 3-26　"按钮 2"的定位

约束布局运行结果如图 3-27 所示。

设计图与运行图相比，两个按钮的显示位置还是有差异，这是因为设计图并不是按实际设备的屏幕分辨率来设定的，距离真正的所见即所得还有一定的差距。当控件定位属性不全时，组件树会有红色圆形警告标识 ❶，Code 视图的控件标签也会标红，同时会多出一个前缀为"tools:"的属性。例如，tools:layout_editor_absoluteX="180dp"指明当前控件在 Design 视图中的 X 轴坐标是 180dp。此属性只在 Design 视图中起辅助定位时使用，在运行时还是被忽略的，即运行时控件在 X 轴坐标还是 0（对齐屏幕左侧）。

如果对"按钮 1"在水平方向分别将左边缘和右边缘依次定位到屏幕两侧，Android Studio 会将"按钮 1"在水平方向自动置中，定位标尺直线变为折线，其含义是最终定位还需考虑其他定位属性。Design 视图显示水平方向双重定位，如图 3-28 所示。运行程序时，"按钮 1"也会显示在屏幕水平方向正中央。

图 3-27　约束布局运行结果

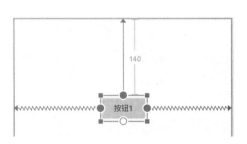

图 3-28　水平方向双重定位

此时"按钮 1"隐含以下属性(此时布局文件中不会显示此属性)：

```
app:layout_constraintHorizontal_bias = "0.5"
```

只要将"按钮 1"水平拖动,如图 3-29 所示,将多出属性 layout_constraintHorizontal_bias,其代表左右定位尺寸(左右圆形句柄到定位基线的距离,定位基线可能是屏幕边缘,也可能是其他控件边缘)的偏离率,0.5 代表按钮左右定位长度相等,小于 0.5 时按钮偏向左边,大于 0.5 时按钮偏向右边。可以在属性栏中直接修改偏离率,也可以在 Inspector(layout_constraintHorizontal_bias 属性,如图 3-29 标注 Constraint Widget 的图形部分)中直接拖动带数字的进度条修改偏离率,还可以通过直接拖动"按钮 1"的方式或在 Code 视图中修改偏离率。

如果希望"按钮 1"无论怎么左右移动,按钮左侧到屏幕左边缘均至少保留 50dp 的间隙,可先选中"按钮 1",然后在 Inspector 左侧文本框中输入 50,layout_marginStart 属性如图 3-30 所示。"按钮 1"左侧折线连接一段 50dp 的直线,相应的 layout_marginStart 属性为 50dp,此时偏离率是不计算这 50dp 直线长度的,即使 app: layout_constraintHorizontal_bias 属性值等于 0,"按钮 1"左侧还是会保留 50dp 的空白,此时拖动

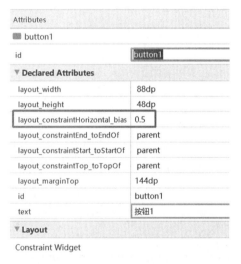

图 3-29　layout_constraintHorizontal_bias 属性

"按钮1"到50dp时就无法再向左边移动。

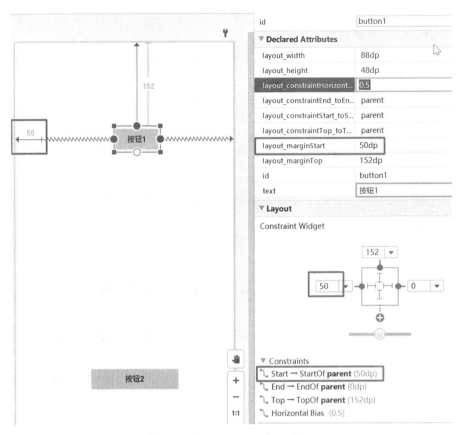

图 3-30　layout_marginStart 属性

选中控件的圆形句柄,按 delete 键可以删除选中的定位。如果选中的是控件,按 delete 键会删除整个控件(含控件定位属性)。

Inspector 如图 3-31 所示。

>> 表示 wrap_content,Code 视图中的属性为 android:layout_width="wrap_content"。

|—| 表示固定值,给控件指定了一个固定的长度或者宽度值。

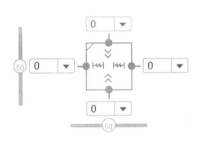

图 3-31　Inspector

|⋀⋀| 表示任意长度或者宽度值。以控件宽度为例,Code 视图中控件的宽度属性变为 android:layout_width="0"。配合其他定位属性,控件宽度可能为 wrap_content(左右圆形句柄只有一个用于定位)、match_parent(左右圆形句柄都用于定位)或任意长度(左右圆形句柄都用于定位,且同时定义了 layout_marginStart 或 layout_marginEnd)。

在"按钮1"上右击,弹出控件快捷菜单,如图 3-32 所示。

选择 Show Baseline 菜单,会在"按钮1"上显示 Baseline(即基准线),如图 3-33 所示。单击"按钮1"的基准线并拖放到"按钮2"的基准线位置就可实现基准线对齐。基准线对齐主要用在多个高度不同的控件间实现文字对齐(如果按钮中文字显示为多行或两个按钮的字体大小不同,基准线对齐是将第一行文字底部对齐)。

图 3-32 控件快捷菜单　　　　　图 3-33 Baseline

选择 Clear Constraints of Selection 菜单,将删除选中控件的所有约束布局属性。

选择 Convert view 菜单,弹出转换控件窗口,如图 3-34 所示。选择想变更的控件类型(Android 中称之为 View),可以在保留定位数据的情况下变更控件类型。

其他菜单项是常用选项,如复制、粘贴等,还有几个菜单项在设计图上方的工具栏中有相同功能按钮。相关功能可参看后续内容。

Transforms 如图 3-35 所示。在 Transforms 中可设置 View 的 X、Y、Z 轴的旋转和坐标参照点的偏移。

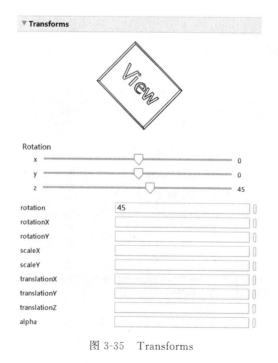

图 3-34 转换控件窗口　　　　　图 3-35 Transforms

图 3-35 中对 Z 轴旋转了 45°,对应属性 android:rotation="45"。如果定义在按钮控件内,则对应按钮控件旋转 45°。如果定义在 ConstraintLayout 标签内,则 ConstraintLayout 标签内的所有控件都会旋转 45°。

```xml
【main.xml】
01  <?xml version = "1.0" encoding = "UTF-8"?>
02  <androidx.constraintlayout.widget.ConstraintLayout xmlns:android = "http://schemas.android.com/apk/res/android"
03      xmlns:app = "http://schemas.android.com/apk/res-auto"
04      xmlns:tools = "http://schemas.android.com/tools"
05      android:layout_width = "match_parent"
06      android:layout_height = "match_parent"
07      android:rotation = "-45">
08
09      <Button
10          android:id = "@+id/button1"
11          android:layout_width = "88dp"
12          android:layout_height = "wrap_content"
13          android:layout_marginStart = "196dp"
14          android:layout_marginTop = "164dp"
15          android:text = "随布局旋转按钮 1"
16          app:layout_constraintStart_toStartOf = "parent"
17          app:layout_constraintTop_toTopOf = "parent" />
18
19      <Button
20          android:id = "@+id/button2"
21          android:layout_width = "88dp"
22          android:layout_height = "wrap_content"
23          android:layout_marginStart = "108dp"
24          android:layout_marginTop = "288dp"
25          android:rotation = "90"
26          android:text = "自定义旋转的按钮 2"
27          app:layout_constraintStart_toStartOf = "parent"
28          app:layout_constraintTop_toTopOf = "parent" />
29
30  </androidx.constraintlayout.widget.ConstraintLayout>
```

第 7 行定义约束布局内所有控件都旋转 -45°。此时 button1 和 button2 都旋转 -45°。第 25 行定义 button2 旋转 90°,扣除约束布局旋转的 -45°,最终效果是 button2 旋转 45°。

android:rotation 属性运行结果如图 3-36 所示。其前缀是 android,属于 android 命名空间,所以也可以用在其他布局中,如线性布局。android:rotation 是以控件中心且垂直于屏幕为轴心的 Z 轴旋转。android:rotationX 和 android:rotationY 分别对应控件 X 方向中心轴和 Y 方向中心轴旋转。

视频讲解

3.10.2 Barrier

在实际布局中可能会遇到 Barrier 定位,如图 3-37 所示。希望"按钮 3"布置在"按钮 1"和"按钮 2"最右侧 40dp 的位置,即如果"按钮 2"比"按钮 1"更靠右,则"按钮 3"左侧距离"按钮 2"右侧 40dp;如果"按钮 1"比"按钮 2"更靠右,则"按钮 3"左侧距离"按钮 1"右侧 40dp。

图 3-36　android:rotation 属性运行结果

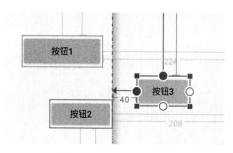

图 3-37　Barrier 定位

为此约束布局引入了 Barrier 的概念,增加了 Barrier 标签,其中定义以下两个属性:

```
app:barrierDirection = "right"
app:constraint_referenced_ids = "button1,button2"
```

以上两行是在"按钮 1"与"按钮 2"右侧建立 Barrier,Barrier 类似一堵墙,两个按钮谁更靠右,这堵墙就以谁为边界。而"按钮 3"左侧定位以这个墙为基准。使用图形化界面建立 Barrier 方法如下:建立"按钮 1"与"按钮 2"并完成相应定位。选中"按钮 1",单击 按钮,弹出如图 3-38 所示的界面,选择 Add Vertical Barrier 菜单添加一个垂直方向 Barrier。

默认添加的 Barrier 与"按钮 1"左侧对齐。修改 Barrier 属性,如图 3-39 所示,在 barrier 属性栏中修改 barrierDirection 属性为 right(或者是 end)、constraint_referenced_ids 属性为 "button1,button2"(默认只有 button1)。如果事先已经明确"按钮 1"和"按钮 2"共同建立 Barrier,也可同时选中"按钮 1"和"按钮 2",然后再添加垂直方向的 Barrier,constraint_referenced_ids 属性自动填写为"button1,button2"。

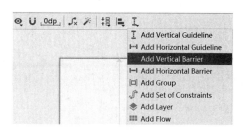

图 3-38　添加垂直方向 Barrier

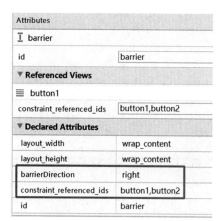

图 3-39　修改 barrier 属性

选中"按钮 3",添加左侧定位到任意控件右边缘,使"按钮 3"的 Declared Attributes 属性栏中多出一项 layout_constraintStart_toEndOf 属性(也可以在 All Attributes 中查找对应属性),将其改为要对齐的 Barrier,将 layout_marginStart 改为 40dp。Android Studio 自动生成布局文件,代码如下:

```
【main.xml】
01  <?xml version = "1.0" encoding = "UTF-8"?>
02  < androidx.constraintlayout.widget.ConstraintLayout xmlns:android = "http://schemas.android.com/apk/res/android"
03      xmlns:app = "http://schemas.android.com/apk/res-auto"
04      xmlns:tools = "http://schemas.android.com/tools"
05      android:layout_width = "match_parent"
06      android:layout_height = "match_parent">
07
08      < Button
09          android:id = "@+id/button1"
10          android:layout_width = "132dp"
11          android:layout_height = "50dp"
12          android:layout_marginTop = "176dp"
13          android:layout_marginEnd = "224dp"
14          android:text = "按钮 1"
15          app:layout_constraintEnd_toEndOf = "parent"
16          app:layout_constraintTop_toTopOf = "parent" />
17
18      < Button
19          android:id = "@+id/button2"
20          android:layout_width = "103dp"
21          android:layout_height = "45dp"
22          android:layout_marginEnd = "208dp"
23          android:layout_marginBottom = "400dp"
24          android:text = "按钮 2"
25          app:layout_constraintBottom_toBottomOf = "parent"
26          app:layout_constraintEnd_toEndOf = "parent" />
27
28      < Button
29          android:id = "@+id/button3"
30          android:layout_width = "wrap_content"
31          android:layout_height = "wrap_content"
32          android:layout_marginStart = "40dp"
33          android:layout_marginTop = "236dp"
34          android:text = "按钮 3"
35          app:layout_constraintStart_toEndOf = "@id/barrier1"
36          app:layout_constraintTop_toTopOf = "parent" />
37
38      < androidx.constraintlayout.widget.Barrier
```

```
39        android:id = "@ + id/barrier1"
40        android:layout_width = "wrap_content"
41        android:layout_height = "wrap_content"
42        app:barrierDirection = "right"
43        app:constraint_referenced_ids = "button1,button2"
44        tools:layout_editor_absoluteX = "55dp" />
45
46  </androidx.constraintlayout.widget.ConstraintLayout>
```

第35行定义button3左侧对齐到barrier1右侧。

第38～44行定义barrier1，其中第43行定义barrier1阻挡的控件是button1和button2，第42行指明barrier1阻挡方向是右侧。

3.10.3 Guideline

视频讲解

之前的控件定位都是基于屏幕（更准确的称呼为控件父容器的约束布局）或者是可见控件，约束布局中引入了一种在运行时看不见的Guideline——定位基准线作为定位补充。添加Guideline如图3-40所示，分别添加垂直和水平方向的Guideline。

拖动Guideline至所需位置，指定左定位150dp的垂直方向Guideline，如图3-41所示。

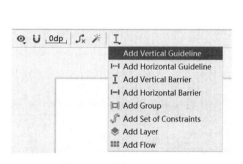

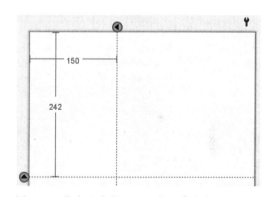

图3-40　添加Guideline　　　　图3-41　指定左定位150dp的垂直方向Guideline

单击左侧向上箭头将切换为向下箭头，此时Guideline按Bottom位置定位，再次单击将切换成百分比符号，代表Guideline使用位置百分比设置自身定位（Guideline属性app：layout_constraintGuide_percent="0.33"是将Guideline设置在屏幕长度或宽度的1/3位置），指定百分比的水平方向Guideline，如图3-42所示。

【注】 目前约束布局版本水平方向的Guideline可通过鼠标单击实现 ▲、▼、% 三种定位方式循环切换，垂直方向的Guideline的切换还有问题。约束布局的功能在不停地升级，或许下一版本会将这个问题解决。

添加控件并将其定位指向Guideline，使用Guideline定位如图3-43所示。实际运行时Guideline是不可见的。

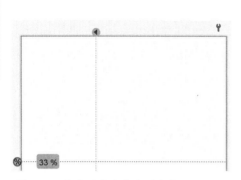
图 3-42 指定百分比的水平方向 Guideline

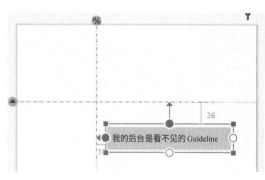
图 3-43 使用 Guideline 定位

布局代码如下:

```
【main.xml】
01  <?xml version = "1.0" encoding = "UTF - 8"?>
02  < androidx.constraintlayout.widget.ConstraintLayoutxmlns:android = "http://schemas.
    android.com/apk/res/android"
03      xmlns:app = "http://schemas.android.com/apk/res - auto"
04      xmlns:tools = "http://schemas.android.com/tools"
05      android:layout_width = "match_parent"
06      android:layout_height = "match_parent">
07
08      < androidx.constraintlayout.widget.Guideline
09          android:id = "@ + id/vertical_guide_line"
10          android:layout_width = "wrap_content"
11          android:layout_height = "wrap_content"
12          android:orientation = "vertical"
13          app:layout_constraintGuide_begin = "150dp" />
14
15      < androidx.constraintlayout.widget.Guideline
16          android:id = "@ + id/percent_guide_Line"
17          android:layout_width = "wrap_content"
18          android:layout_height = "wrap_content"
19          android:orientation = "horizontal"
20          app:layout_constraintGuide_percent = "0.33" />
21
22      < Button
23          android:layout_width = "wrap_content"
24          android:layout_height = "wrap_content"
25          android:layout_marginStart = "16dp"
26          android:layout_marginTop = "36dp"
27          android:text = "我的后台是看不见的 Guideline"
28          app:layout_constraintStart_toStartOf = "@ id/vertical_guide_line"
29          app:layout_constraintTop_toTopOf = "@ + id/percent_guide_Line" />
30
31  </androidx.constraintlayout.widget.ConstraintLayout >
```

第 8~13 行定义了一个距离屏幕左边界 150dp 的垂直方向 Guideline。

第15~20行定义了一个距离屏幕上端1/3位置的水平方向Guideline。

第22~29行定义了一个按钮,其顶端和左侧分别定位到两个Guideline。由于Guideline是一条线,因此第28行layout_constraintStart_toStartOf换成layout_constraintStart_toEndOf的效果是一样的。

3.10.4 Group

使用约束布局的一个目的是减少布局的嵌套。如果想把多个控件设置为隐藏,就需分别设置各控件的可视化属性。使用Group相当于将各控件进行分组,设置Group属性等效于将Group中各控件设置相同属性。在Design视图中添加Group时需先选择要包含的控件,然后如图3-44所示选择Add Group菜单项添加的Group对象,新添加Group对象取名为group1。

group1在程序运行时是不可见的,要将相应的控件放入group1,最便捷的方式是在组件树中将相应的按钮控件拖放到group1,如图3-45所示。

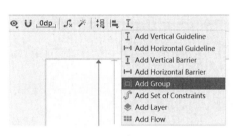

图3-44 添加Group　　　　　图3-45 将按钮添加到group1

此时group1中包含了button1和button2。Group的关键代码如下:

```
01    <androidx.constraintlayout.widget.Group
02        android:id = "@+id/group1"
03        android:layout_width = "wrap_content"
04        android:layout_height = "wrap_content"
05        app:constraint_referenced_ids = "button1,button2" />
```

其中,第5行定义group1中包含button1和button2。运行程序,可以看到屏幕上显示button1和button2。改变group1的visibility属性为invisible,再次运行程序,屏幕上不显示button1和button2。需要注意的是,group1的属性栏中有两个visibility属性,如图3-46所示。

图3-46 visibility属性

上方的visibility属性在Code视图中显示为android:visibility,其设定的值影响相关控件在Android设备上是否显示。下方的visibility属性前有一个🔧符号,在Code视图中显示为 tools:visibility,其设定只影响在 Android Studio 的 Design 视图中是否显示,并不影响在 Android 设备上的运行显示。

3.10.5 Circle

Circle 方式定位目前还无法使用图形界面操作,可以在 Code 视图中输入代码。Circle 定位示意图如图 3-47 所示。以参考定位控件 A 中心为圆点,以控件 B 中心到控件 A 中心为半径,与垂直向上方向的夹角来定位控件 B 的位置。

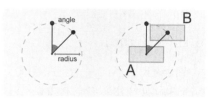

图 3-47　Circle 定位示意图

```
01  <?xml version = "1.0" encoding = "UTF - 8"?>
02  <androidx.constraintlayout.widget.ConstraintLayout xmlns:android = "http://schemas.
    android.com/apk/res/android"
03      xmlns:app = "http://schemas.android.com/apk/res - auto"
04      xmlns:tools = "http://schemas.android.com/tools"
05      android:layout_width = "match_parent"
06      android:layout_height = "match_parent">
07
08      <Button
09          android:id = "@ + id/button1"
10          android:layout_width = "132dp"
11          android:layout_height = "50dp"
12          android:layout_marginStart = "84dp"
13          android:layout_marginTop = "160dp"
14          android:text = "按钮 1"
15          app:layout_constraintStart_toStartOf = "parent"
16          app:layout_constraintTop_toTopOf = "parent" />
17
18      <Button
19          android:id = "@ + id/button2"
20          android:layout_width = "103dp"
21          android:layout_height = "45dp"
22          android:text = "按钮 2"
23          app:layout_constraintCircle = "@id/button1"
24          app:layout_constraintCircleAngle = "45"
25          app:layout_constraintCircleRadius = "150dp" />
26
27  </androidx.constraintlayout.widget.ConstraintLayout>
```

第 23 行定义 button2 按照 Circle 方式来定位,定位基准为 button1。

第 24 行定义夹角为 45°。

第 25 行定义 button1 与 button2 中心点的距离为 150dp。

目前,Circle 方定位方式还不完善,除了不支持图形化拖曳设计以外,在 Code 视图下 Button 标签也会显示为红色,组件树中 button2 也会用红色惊叹号标识,提示相关约束属性不完备。

3.10.6　Chain

Chain 用于指定链式约束。Chain 示意图如图 3-48 所示。图 3-48 中，控件 A 与控件 B 相互约束形成一个简单的链式约束。

链式约束中的第一个控件称为 Chain Head。在整个约束链中只要在 Chain Head 中定义链式约束类型即可。Chain Head 示意图如图 3-49 所示。

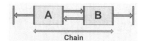

图 3-48　Chain 示意图

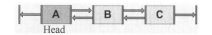

图 3-49　Chain Head 示意图

在 Head 控件中添加 layout_constraintHorizontal_chainStyle 属性，ChainStyle 属性及示意图如图 3-50 所示。5 种链式约束类型的主要区别是控件间的间隔或控件自身宽度比例。如果 Head 控件中未设置 layout_constraintHorizontal_chainStyle 属性，则默认为 Spread Chain 类型。

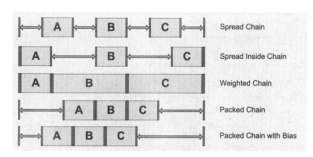

图 3-50　ChainStyle 属性及示意图

以下是根据官方代码编写的案例，分别实现图 3-50 的 5 种链式约束类型。

【main.xml】
```
01  <?xml version = "1.0" encoding = "UTF - 8"?>
02  < androidx.constraintlayout.widget.ConstraintLayout xmlns:android = "http://schemas.
    android.com/apk/res/android"
03      xmlns:app = "http://schemas.android.com/apk/res - auto"
04      android:layout_width = "match_parent"
05      android:layout_height = "match_parent">
06
07      <!-- 默认为 app:layout_constraintVertical_chainStyle = "spread" -->
08      < Button
09          android:id = "@ + id/spread1"
10          android:layout_width = "wrap_content"
11          android:layout_height = "wrap_content"
12          android:text = "1"
13          app:layout_constraintLeft_toLeftOf = "parent"
14          app:layout_constraintRight_toLeftOf = "@ + id/spread2"
15          app:layout_constraintTop_toTopOf = "parent" />
16
```

```xml
17      <Button
18          android:id = "@ + id/spread2"
19          android:layout_width = "wrap_content"
20          android:layout_height = "wrap_content"
21          android:text = "2"
22          app:layout_constraintLeft_toRightOf = "@ + id/spread1"
23          app:layout_constraintRight_toLeftOf = "@ + id/spread3"
24          app:layout_constraintTop_toTopOf = "@ + id/spread1" />
25
26      <Button
27          android:id = "@ + id/spread3"
28          android:layout_width = "wrap_content"
29          android:layout_height = "wrap_content"
30          android:text = "3"
31          app:layout_constraintLeft_toRightOf = "@ + id/spread2"
32          app:layout_constraintRight_toRightOf = "parent"
33          app:layout_constraintTop_toTopOf = "@ + id/spread1" />
34
35      <!-- app:layout_constraintHorizontal_chainStyle = "spread_inside" -->
36      <Button
37          android:id = "@ + id/spread_inside1"
38          android:layout_width = "wrap_content"
39          android:layout_height = "wrap_content"
40          android:text = "spread_in1"
41          app:layout_constraintHorizontal_chainStyle = "spread_inside"
42          app:layout_constraintLeft_toLeftOf = "parent"
43          app:layout_constraintRight_toLeftOf = "@ + id/spread_inside2"
44          app:layout_constraintTop_toBottomOf = "@ + id/spread1" />
45
46      <Button
47          android:id = "@ + id/spread_inside2"
48          android:layout_width = "wrap_content"
49          android:layout_height = "wrap_content"
50          android:text = "spread_in2"
51          app:layout_constraintLeft_toRightOf = "@ + id/spread_inside1"
52          app:layout_constraintRight_toLeftOf = "@ + id/spread_inside3"
53          app:layout_constraintTop_toTopOf = "@ + id/spread_inside1" />
54
55      <Button
56          android:id = "@ + id/spread_inside3"
57          android:layout_width = "wrap_content"
58          android:layout_height = "wrap_content"
59          android:text = "spread_in3"
60          app:layout_constraintLeft_toRightOf = "@ + id/spread_inside2"
61          app:layout_constraintRight_toRightOf = "parent"
62          app:layout_constraintTop_toTopOf = "@ + id/spread_inside1" />
63
64      <!-- app:layout_constraintHorizontal_weight 设置各控件比例 -->
65      <Button
66          android:id = "@ + id/weight1"
67          android:layout_width = "wrap_content"
68          android:layout_height = "wrap_content"
69          android:text = "weight1"
```

```xml
70          app:layout_constraintLeft_toLeftOf = "parent"
71          app:layout_constraintRight_toLeftOf = "@ + id/weight2"
72          app:layout_constraintTop_toBottomOf = "@ + id/spread_inside1" />
73
74      < Button
75          android:id = "@ + id/weight2"
76          android:layout_width = "0dp"
77          android:layout_height = "wrap_content"
78          android:text = "weight2"
79          app:layout_constraintHorizontal_weight = "2"
80          app:layout_constraintLeft_toRightOf = "@ + id/weight1"
81          app:layout_constraintRight_toLeftOf = "@ + id/weight3"
82          app:layout_constraintTop_toTopOf = "@ + id/weight1" />
83
84      < Button
85          android:id = "@ + id/weight3"
86          android:layout_width = "0dp"
87          android:layout_height = "wrap_content"
88          android:text = "weight3"
89          app:layout_constraintHorizontal_weight = "3"
90          app:layout_constraintLeft_toRightOf = "@ + id/weight2"
91          app:layout_constraintRight_toRightOf = "parent"
92          app:layout_constraintTop_toTopOf = "@ + id/weight1" />
93
94
95      <!-- app:layout_constraintHorizontal_chainStyle = "packed" -->
96      < Button
97          android:id = "@ + id/packed1"
98          android:layout_width = "wrap_content"
99          android:layout_height = "wrap_content"
100         android:text = "packed1"
101         app:layout_constraintHorizontal_chainStyle = "packed"
102         app:layout_constraintLeft_toLeftOf = "parent"
103         app:layout_constraintRight_toLeftOf = "@ + id/packed2"
104         app:layout_constraintTop_toBottomOf = "@ + id/weight1" />
105
106     < Button
107         android:id = "@ + id/packed2"
108         android:layout_width = "wrap_content"
109         android:layout_height = "wrap_content"
110         android:text = "packed2"
111         app:layout_constraintLeft_toRightOf = "@ + id/packed1"
112         app:layout_constraintRight_toLeftOf = "@ + id/packed3"
113         app:layout_constraintTop_toTopOf = "@ + id/packed1" />
114
115     < Button
116         android:id = "@ + id/packed3"
117         android:layout_width = "wrap_content"
118         android:layout_height = "wrap_content"
119         android:text = "packed3"
120         app:layout_constraintLeft_toRightOf = "@ + id/packed2"
121         app:layout_constraintRight_toRightOf = "parent"
122         app:layout_constraintTop_toTopOf = "@ + id/packed1" />
```

```
123
124        <!-- app:layout_constraintHorizontal_bias = "0.2" -->
125        <!-- app:layout_constraintHorizontal_chainStyle = "packed" -->
126        < Button
127            android:id = "@ + id/bias1"
128            android:layout_width = "wrap_content"
129            android:layout_height = "wrap_content"
130            android:text = "bias1"
131            app:layout_constraintHorizontal_bias = "0.2"
132            app:layout_constraintHorizontal_chainStyle = "packed"
133            app:layout_constraintLeft_toLeftOf = "parent"
134            app:layout_constraintRight_toLeftOf = "@ + id/bias2"
135            app:layout_constraintTop_toBottomOf = "@ + id/packed1" />
136
137        < Button
138            android:id = "@ + id/bias2"
139            android:layout_width = "wrap_content"
140            android:layout_height = "wrap_content"
141            android:text = "bias2"
142            app:layout_constraintLeft_toRightOf = "@ + id/bias1"
143            app:layout_constraintRight_toLeftOf = "@ + id/bias3"
144            app:layout_constraintTop_toTopOf = "@ + id/bias1" />
145
146        < Button
147            android:id = "@ + id/bias3"
148            android:layout_width = "wrap_content"
149            android:layout_height = "wrap_content"
150            android:text = "bias3"
151            app:layout_constraintLeft_toRightOf = "@ + id/bias2"
152            app:layout_constraintRight_toRightOf = "parent"
153            app:layout_constraintTop_toTopOf = "@ + id/bias1" />
154
155    </androidx.constraintlayout.widget.ConstraintLayout>
```

5 种链式约束类型运行结果如图 3-51 所示。

图 3-51　5 种链式约束类型运行结果

在 Code 视图中按钮的文本是小写,但在 Design 视图和运行结果中都是大写,这是因为从 Android 5.0 起按钮的属性默认为 android:textAllCaps="true"。只要在按钮属性中增加 android:textAllCaps="false"即可恢复原有的大小写显示。相应的 Java 命令为 button.setAllCaps(false)。

3.11　Space 和 layout_margin

在布局中有时需要将 View 与其他 View 保持一定的间隔,可以使用 Space 控件来填充 View 之间的间隔。Space 控件在运行时是不可见的。也可以使用 View 的 layout_margin、layout_marginEnd 等属性来设定 View 与其他 View 的间隔。

【main.xml】
```
01  <?xml version = "1.0" encoding = "UTF - 8"?>
02  < LinearLayout xmlns:android = "http://schemas.android.com/apk/res/android"
03
04      android:layout_width = "match_parent"
05      android:layout_height = "match_parent"
06      android:orientation = "vertical">
07
08      < Button
09          android:id = "@ + id/button1"
10          android:layout_width = "match_parent"
11          android:layout_height = "wrap_content"
12          android:text = "织女" />
13
14      <!-- 占据一定的空间,但是却不显示任何东西 -->
15
16      < Space
17          android:layout_width = "match_parent"
18          android:layout_height = "100dp" />
19
20      < Button
21          android:id = "@ + id/button2"
22          android:layout_width = "match_parent"
23          android:layout_height = "wrap_content"
24          android:textAllCaps = "false"
25          android:text = "牛郎与织女隔着100dp的 Space" />
26
27      < Button
28          android:id = "@ + id/button3"
29          android:layout_width = "match_parent"
30          android:layout_height = "wrap_content"
31          android:text = "参照基准" />
32
33      < Button
34          android:id = "@ + id/button4"
```

```
35          android:layout_width = "match_parent"
36          android:layout_height = "wrap_content"
37          android:layout_margin = "20dp"
38          android:textAllCaps = "false"
39          android:text = "layout_margin = '20dp'上下左右都留 20dp 空白" />
40
41      < Button
42          android:id = "@ + id/button5"
43          android:layout_width = "match_parent"
44          android:layout_height = "wrap_content"
45          android:text = "参照基准" />
46
47      < Button
48          android:id = "@ + id/button6"
49          android:layout_width = "match_parent"
50          android:layout_height = "wrap_content"
51          android:layout_marginEnd = "20dp"
52          android:textAllCaps = "false"
53          android:text = "layout_marginRight = '20dp'右边留 20dp 空白" />
54  </LinearLayout >
```

第 16～18 行定义控件 Space，其高度为 100dp。Space 位于 button1 与 button2 之间，所以两个按钮在垂直方向上间隔 100dp 的距离。

第 37 行定义 button4 在上下左右 4 个方向都保留 20dp 的空白。

将第 35 行的 match_parent 换成 wrap_content，注意观察 button4 位置的变化。

Space 和 layout_margin 运行结果如图 3-52 所示。

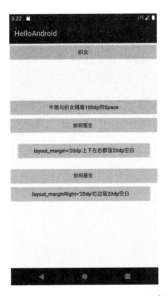

图 3-52　Space 和 layout_margin 运行结果

第 4 章　Android 常用控件

Android 定义了一个 View 类，是所有控件的基础类。Activity 调用 Window，再通过 Window 来展现 View。具体如何定位 View 可参见第 3 章的布局来确定。本章开始讲解各类常用控件的属性、方法和使用注意事项。在布局文件中可以定义控件的属性，在 Java 文件的 Activity 中使用方法来变更控件的属性或执行特定的操作。

4.1　TextView

TextView(文本框)控件(以下简称 TextView，其余控件与此同)的作用是在界面上显示文字。用向导生成的第一个程序在屏幕正中显示的"Hello World!"就是 TextView。TextView 只能显示文字，不能使用软键盘输入文字。TextView 又派生出了响应用户文字输入的 EditText 和响应用户单击操作的 Button 等控件。下面通过案例来了解 TextView 的常用属性和方法。

4.1.1　TextView 的常用属性和方法

视频讲解

```
【main.xml】
01  <?xml version = "1.0" encoding = "UTF - 8"?>
02  < LinearLayout xmlns:android = "http://schemas.android.com/apk/res/android"
03      android:layout_width = "match_parent"
04      android:layout_height = "match_parent"
05      android:orientation = "vertical">
06
07      < TextView
08          android:id = "@ + id/textView1"
09          android:layout_width = "match_parent"
10          android:layout_height = "wrap_content"
11          android:text = "@string/hello" />
12
13  </LinearLayout>
```

在一个垂直线性布局中定义了一个 TextView。TextView 的属性除了在布局文件中定义以外，也可以在 Java 文件的 Activity 中设置。两种方式各有优劣，可以根据具体情况和个人喜好选择使用。

【FirstActivity.java】
```
01    public class FirstActivity extends Activity
02    {
03        @Override
04        public void onCreate(Bundle savedInstanceState)
05        {
06            super.onCreate(savedInstanceState);
07            setContentView(R.layout.main);
08
09            TextView textView1 = (TextView) findViewById(R.id.textView1);
10            textView1.setText("1234567890\n11111111.1\nabcdefghij\nABCDEFGHIJ\n 一二三四五六七八九十零");
11            textView1.setTextSize(20);                                    //设置字体大小
12            textView1.setTextColor(Color.RED);                            //设置字体颜色
13            textView1.setBackgroundColor(Color.rgb(220, 220, 220));       //设置背景颜色
14            textView1.append("\n      本行是 append 添加的文字 1234567890");
15            textView1.append("\n\n 以下内容是从 TextView 对象中获取值：\n" + textView1.getText());
16        }
17    }
```

回顾前面的知识，第1行是系统自动生成的代码，定义了一个继承于 Activity 的公共类 FirstActivity。

第3行是伪代码，指明第4行的 onCreate() 方法是重写父类的 onCreate() 方法。

第7行加载 main.xml 布局文件并显示到屏幕。

第9行定义一个 TextView 类型变量 textView1。R.id.textView1 对应布局文件 main.xml 中第8行的 id。这里使用了 findViewById() 方法，此方法的作用是通过整型资源变量 R.id.textView1 查找并返回对应的 View，然后通过强制类型转换（TextView）将 View 转换为 TextView 类型对象。等号右边的 textView1 代表的是控件 id 的整型值变量，等号左边的 textView1 代表 TextView 类型对象变量，两者是不同的概念。此时对等号左边 textView1 的操作就是对布局文件中定义的 TextView 控件进行操作。为了叙述方便，后面案例如果存在声明的控件变量名与布局文件中控件 id 名相同时，所列名称默认是控件对象变量名。

第10行的 setText() 方法是将参数字符串覆盖 textView1 的原有值（布局文件第11行定义的值）。字符串"\n"的含义与 Java 中相同，代表换行。

第11行的 setTextSize() 方法设置 textView1 的字体大小，默认单位是 sp。

第12行的 setTextColor() 方法设置字体颜色，Color.RED 是内置的整型常量，代表红色。

第13行的 setBackgroundColor() 方法设置字体背景色，使用 Color.rgb 方法设置颜色。

设置颜色的常用方式有如下4种。

（1）在布局文件中使用十六进制数字，如 android:textColor="#f0f0f0"，红、绿、蓝三色各占十六进制数的两位。

(2) 在布局文件中使用系统定义颜色,如 android:textColor = "@android:color/black"。

(3) 在 Java 文件中使用 Color 类中已定义的整型常量,如 Color.RED。常量名一般大写。

(4) 在 Java 文件中使用 Color 类中的 rgb()方法,如 Color.rgb(220,220,220),3 个参数分别代表红、绿、蓝三色。在 Android Studio 中参数提示如图 4-1 所示。

其中,灰色显示字体如"red:"代表方法的第一个形参名称,只起到提示的作用,在实际编写的代码中并不存在。

第 14 行的 append()方法的作用是将字符串附加在 textView1 已有字符串之后。

第 15 行调用 getText()方法返回 textView1 的当前字符串值,之后用 append()方法将返回值附加在 textView1 已有字符串之后。

TextView 运行结果如图 4-2 所示。

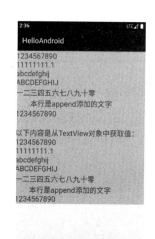

Color.rgb(red: 220, green: 220, blue: 220)

图 4-1　参数提示　　　　　　图 4-2　TextView 运行结果

布局文件的第 10 行定义 textView1 的高度为 wrap_content,实际显示的高度以容纳 textView1 的字符串内容为准。如果字符串内容超过父容器高度则以父容器高度为准,超出的内容是无法通过拉动屏幕来查看的,要实现类似功能需要将 textView1 放置到 ScrollView 中,相关内容在后面章节详述。

另一个要注意的问题是文字的宽度。图 4-2 中第 2 行显示的数字由于加入了小数点,同样位数的数字(含小数点)第 1 行的显示宽度要更大(前 8 个数字一样宽,到小数点才有变化)。同样数量的大小写字母显示的宽度也不一样。中文字符显示得要更宽一些。要保证所有数字和字母都对齐,可以选择 Android 系统默认字体为等宽字体,或者使用以下命令:

`textView1.setTypeface(Typeface.MONOSPACE);`

setTypeface()方法可以设置字体和类型,Typeface.MONOSPACE 是将字体设为等宽字体,设定 Typeface.MONOSPACE 效果如图 4-3 所示。除中文以外,其他字符都是按相同宽度显示的。此方法还可以设置斜体(Typeface.ITALIC)、粗体(Typeface.DEFAULT_

BOLD)等。如果同时设置多个属性(如设置等宽、加粗字体)可使用以下命令:

```
textView1.setTypeface(Typeface.MONOSPACE, Typeface.BOLD);
```

如果添加代码"**textView1.setGravity(Gravity.RIGHT);**",则将textView1内的文字设为右对齐,等效于布局文件中的 **android:gravity="right"**。设定setGravity()右对齐,运行结果如图4-4所示。

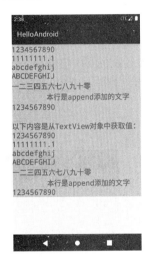 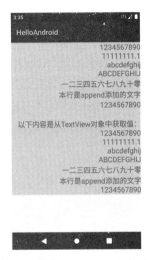

图4-3 设定Typeface.MONOSPACE效果   图4-4 设定setGravity()右对齐

如果添加代码"**textView1.setHeight(500);**",则无视textView1的文本内容长度,将textView1的高度固定为500px。设定setHeight(),运行结果如图4-5所示。

如果添加代码"**textView1.setMaxLines(3);**",则textView1的文字最多显示3行。设定setMaxLines(),运行结果如图4-6所示。如果设置的行数超过实际文本显示的行数,则以实际文本行数显示。

图4-5 设定setHeight()   图4-6 设定setMaxLines()

当 setHeight()与 setMaxLines()方法同时使用时,如果 setMaxLines()方法设置的行数高度小于 setHeight()指定的高度,则显示 setMaxLines()方法设置的行数,textView1 的高度由 setHeight()确定。setHeight()与 setMaxLines()同时使用,运行结果如图 4-7 所示。如果 setMaxLines()方法设置的行数高度大于 setHeight()指定的高度,则 textView1 的高度由 setMaxLines()的显示行数(不一定是设定行数,如果显示文本行数小于设定行数,则显示行数为显示文本行数)的高度确定。

如果添加代码"**textView1. setMinLines(2);**",则 textView1 的文本至少显示 2 行,等效于布局文件中的 **android:minLines="2"**。

获取 textView1 的 id 值可使用如下两条代码,运行结果如图 4-8 所示。

```
textView1.setText("textView1 的 R.id.textView1 为: " + R.id.textView1);
textView1.append("\ntextView1 的 getId 为: " + textView1.getId());
```

如果添加代码"**setPadding(100,100,100,100);**",则 textView1 的文本按距离 textView1 的 left、top、right 和 bottom 边缘各 100px 距离显示。设定 setPadding(),运行结果如图 4-9 所示。

图 4-7　setHeight()与 setMaxLines()同时使用　　　图 4-8　获取 id 值　　　图 4-9　设定 setPadding()

4.1.2　theme 和 style

为了保持布局文件有统一风格,可以定义 theme(主题)或 style(风格)并在布局中引用。两者的主要区别:theme 除了能像 style 一样用于控件属性中,还可以用于 AndroidManifest.xml 中的 application 或 activity 标签中,作用于整个应用程序或者 Activity,如 **android:theme="@android:style/Theme.Light"**,可改变标题栏的尺寸和颜色;style 主要作用于具体的 View 控件,如 TextView。当同时定义了 theme 和 style 时,style 的优先级高于 theme。Android 中很多属性的优先级都遵循作用域越小优先级越高的特性。

同样道理,控件中的属性优先级高于 style 中相关属性的优先级。以下的 style.xml 文件预定义了 style 格式供布局文件中的控件调用。如果多个布局文件都要保持统一的风格,使用 style.xml 来定义风格是一种简便的方案,特别是在项目后期要统一修改风格时,只要修改 style.xml 中的相关标签定义就可以了。

【style.xml】
```
01  <?xml version = "1.0" encoding = "UTF-8"?>
02  <resources>
03    <style name = "style01">
04        <item name = "android:textSize">16sp</item>
05        <item name = "android:textColor">#0000FF</item>
06    </style>
07    <style name = "style02">
08        <item name = "android:textSize">30sp</item>
09        <item name = "android:textColor">#000000</item>
10    </style>
11  </resources>
```

第3~6行定义了 style01,设置字体颜色为蓝色,字体大小为 16sp。
第7~10行定义了 style02,设置字体颜色为黑色,字体大小为 30sp。
以下是对 style 格式进行调用的布局文件源码。

【main.xml】
```
01  <?xml version = "1.0" encoding = "UTF-8"?>
02  <LinearLayout xmlns:android = "http://schemas.android.com/apk/res/android"
03      android:layout_width = "match_parent"
04      android:layout_height = "match_parent"
05      android:orientation = "vertical">
06
07      <TextView
08          android:id = "@+id/TextView1"
09          style = "@style/style01"
10          android:layout_width = "wrap_content"
11          android:layout_height = "wrap_content"
12          android:text = "textSize:16sp, textColor:#0000FF" />
13
14      <TextView
15          android:id = "@+id/TextView2"
16          style = "@style/style02"
17          android:layout_width = "wrap_content"
18          android:layout_height = "wrap_content"
19          android:text = "textSize:30sp, textColor:#000000" />
20
21      <TextView
22          android:id = "@+id/TextView3"
23          android:layout_width = "wrap_content"
24          android:layout_height = "wrap_content"
```

```
25            android:text = "@string/textViewTheme"
26            android:theme = "@style/style01" />
27
28      < TextView
29            android:id = "@ + id/TextView4"
30            style = "@style/style02"
31            android:layout_width = "wrap_content"
32            android:layout_height = "wrap_content"
33            android:text = "&quot;style优先级高于theme&quot;"
34            android:theme = "@style/style01" />
35
36      < TextView
37            android:id = "@ + id/TextView5"
38            style = "@style/style02"
39            android:layout_width = "wrap_content"
40            android:layout_height = "wrap_content"
41            android:text = "View控件中定义的属性优先级高于style"
42            android:textColor = "@android:color/holo_red_dark"
43            android:textSize = "40dip"
44  </LinearLayout >
```

布局文件的垂直线性布局中放置了 5 个 TextView，第 1 个 TextView 虽然没有直接定义字体大小和颜色，但在第 9 行定义了@style/style01，使用在 values 目录下 style.xml 文件中定义的 style01 相关属性。同理，第 2 个 TextView 使用了 style02。

新版本的 SDK 可以将 theme 属性像 style 属性一样用于布局文件的控件中。第 26 行使用 theme 属性来调用 style01。注意此时与 style 属性调用的区别：style 属性前没有"android:"修饰，从语法上来说 theme 更规范一些。第 4 个 TextView 演示同时用 style 和 theme 定义控件属性时的优先级。第 5 个 TextView 演示同时定义控件属性和 style 时的优先级。

另外一个需要注意的是 TextView 中输出双引号的问题。布局文件中不支持 Java 中的转义符号"\""，要输出半角的双引号可采用以下 3 种方式。

(1) 如第 25 行引用 strings.xml 文件中定义的字符串资源，其值可以采用转义符号"\""，推荐使用此方式。strings.xml 文件字符串资源如下所示：

< string name = "textViewTheme" > android: theme = \" @ style/style01\"</string >

(2) 使用 XML 的转义字符"""来输出半角双引号，如第 33 行所示。

(3) 在 Java 文件中使用 setText()方法，方法中字符串用 Java 中的转义符号"\""来输出半角双引号。

theme 和 style 运行结果如图 4-10 所示。

图 4-10 theme 和 style 运行结果

4.1.3 layout_gravity 与 gravity

layout_gravity 用于设定当前控件在父容器中的对齐方式。带"layout_"前缀的一般与当前控件的父容器有关,如 layout_width 等。gravity 用于设置当前控件内部包含的控件或文字对齐方式。

视频讲解

```
【main.xml】
01  <?xml version = "1.0" encoding = "UTF-8"?>
02  <LinearLayout xmlns:android = "http://schemas.android.com/apk/res/android"
03      android:id = "@ + id/lineLayout1"
04      android:layout_width = "match_parent"
05      android:layout_height = "match_parent"
06      android:orientation = "vertical">
07
08      <TextView
09          android:id = "@ + id/textView1"
10          android:layout_width = "match_parent"
11          android:layout_height = "wrap_content"
12          android:background = "@android:color/holo_orange_light"
13          android:gravity = "right"
14          android:text = "match_parent + gravity = right"
15          android:textSize = "24sp" />
16      <TextView
17          android:id = "@ + id/textView2"
18          android:layout_width = "wrap_content"
19          android:layout_height = "wrap_content"
20          android:background = "#ffff00"
21          android:gravity = "right"
22          android:text = "wrap_content + gravity = right"
23          android:textSize = "24sp" />
24      <TextView
25          android:layout_width = "wrap_content"
26          android:layout_height = "wrap_content"
27          android:layout_gravity = "right"
28          android:background = "#ffff00"
29          android:text = "wrap_content + layout_gravity = right"
30          android:textSize = "24sp" />
31  </LinearLayout>
```

第 1 个 TextView 在第 10 行定义控件宽度与父容器等宽,而父容器为线性布局,其宽度也与父容器等宽,所以第 1 个 TextView 与屏幕等宽。第 13 行定义 TextView 控件内部对象(对 textView1 而言就是显示的字符串文本)右对齐。

第 2 个 TextView 在第 18 行定义 TextView 宽度以显示文字(含 padding)宽度为准,此时 TextView 宽度与文字宽度相同,所以第 21 行定义文字右对齐是看不出效果的。

第 3 个 TextView 在第 25 行定义 TextView 宽度以显示文字宽度为准,第 27 行执行相对父容器的右对齐,最终效果是整个 TextView 对齐屏幕右边缘。

layout_gravity 与 gravity 运行结果如图 4-11 所示。

以上方式都是通过布局文件属性来设置对齐方式,如果希望在程序运行时才设置对齐方式,可在 Activity 中添加以下命令:

```
TextView textView1 = (TextView) findViewById(R.id.textView1);
textView1.setGravity(Gravity.CENTER);
```

通过 textView1 的 setGravity()方法(等效于布局文件的 android:gravity)设置文字对齐方式为中对齐。由于 Java 中的命令是运行时起效的,因此上述对齐代码将覆盖布局文件中 textView1 的右对齐属性。setGravity()运行结果如图 4-12 所示。

图 4-11　layout_gravity 与 gravity 运行结果　　　　图 4-12　setGravity()运行结果

Java 命令中没有类似于 android:layout_gravity 的命令,但可以调用父容器(案例中是 linearLayout1,系统默认生成的布局是没有 id 的,需要给线性布局添加 id)的 setGravity()方法来实现类似的效果。

【注】用 setGravity()方法与 android:gravity、android:layout_gravity 的区别:

setGravity()是控件内所有对象按指定方式对齐,等效于 android:gravity; android:layout_gravity 是当前控件在父容器中的对齐,并不影响父容器中的其他控件对齐方式。

4.1.4　findViewById()与 viewBinding

要在 Java 中调用布局文件中控件,可以使用 findViewById()方法,但这种方式也常常被开发人员诟病,原因是布局文件中定义的是控件 id 而不是控件名称,要在 Java 文件中使用控件需要先运行一条 findViewById 命令,然后转换为相应控件类型方可使用。为了取名方便,往往将布局文件中的控件 id 和 Java 文件中的对象变量名设为一样的,这又容易造成概念混淆。在新版的 SDK 中引入了视图绑定的概念。要使用视图绑定,需要在 app 目录下的 build.gradle 文件中加入配置 **viewBinding {enabled = true}** 才能激活视图绑定。最新版本的 Android Studio 可将上述配置替换成:

视频讲解

```
buildFeatures {
    dataBinding = true
    viewBinding = true
}
```

【main.xml】
```
01  <?xml version = "1.0" encoding = "UTF-8"?>
02  < LinearLayout xmlns:android = "http://schemas.android.com/apk/res/android"
03      android:layout_width = "match_parent"
04      android:layout_height = "match_parent"
05      android:orientation = "vertical">
06
07      < TextView
08          android:id = "@ + id/textView1"
09          android:layout_width = "match_parent"
10          android:layout_height = "wrap_content"
11          android:text = "布局中定义的第一个 TextView" />
12
13      < TextView
14          android:id = "@ + id/textView2"
15          android:layout_width = "match_parent"
16          android:layout_height = "wrap_content"
17          android:text = "布局中定义的第二个 TextView" />
18
19  </LinearLayout>
```

布局文件在一个垂直线性布局中放置了两个 TextView 控件。

【FirstActivity.java】
```
01  public class FirstActivity extends Activity
02  {
03      @Override
04      public void onCreate(Bundle savedInstanceState)
05      {
06          super.onCreate(savedInstanceState);
07          setContentView(R.layout.main);
08
09          TextView textView1 = (TextView) findViewById(R.id.textView1);
10          textView1.setText("第一次采用 findViewById()方法");
11
12          MainBinding mainBinding = MainBinding.inflate(getLayoutInflater());
13          setContentView(mainBinding.getRoot());
14          mainBinding.textView2.setText("采用 MainBinding");
15
16          textView1 = (TextView) findViewById(R.id.textView1);
17          textView1.setText("第二次采用 findViewById()方法");
18      }
19  }
```

第9~10行是按传统方式设置textView1。如果无后续代码则在textView1上显示字符串"第一次采用findViewById()方法"。

第12行调用MainBinding类,这个类是由视图绑定激活以后自动生成的,命名方式是布局文件名(大驼峰命名法)前缀+Binding。本例布局文件名称为main.xml,所以类名自动定义为MainBinding。Inflate方式是调用布局填充器。此行代码必须在setContentView()方法前才有效。

第13行是将布局填充器中的所有控件通过setContentView()方法绑定到Activity,此命令会覆盖第7行的绑定命令。

第14行可直接使用MainBinding类实例mainBinding来调用textView2,无须单独调用findViewById()方法赋值给textView2。

第16~17行演示使用视图绑定后仍然可以再调用findViewById()方法调用控件对象。

【注】 执行第13行视图绑定后第9行的textView1已被清空,此时textView1显示的是布局文件中设定的值。如果要使用textView1,需重新调用findViewById()方法绑定控件或者通过视图绑定方式调用,如mainBinding.textView2。

findViewById()与viewBinding运行结果如图4-13所示。

需要注意的是,如果没有第16行的命令,textView1将显示为"布局中定义的第一个TextView",因为程序执行到13行时重新绑定布局文件到Activity。此时的Activity中的textView1需要使用findViewById()方法重新指向布局文件中的TextView控件。在实际开发中使用视图绑定是用第12~13行代码直接替代第7行代码以简化控件调用。

视图绑定的优点:

(1) null安全性。由于视图绑定会创建对视图的直接引用,因此不会因无效的视图id导致null异常风险。

(2) 类型安全性。每个绑定类中的类型都具有与其在布局文件中引用的视图匹配的类型,不存在调用findViewById()方法后所需的类强制转换过程。

视图绑定的缺点:

使用向导新建项目时默认不是使用视图绑定,需要手工修改build.gradle和Java两个文件中相应配置和代码。

图4-13　findViewById()与viewBinding运行结果

【注】 MainBinding.inflate()方法必须在setContentView(mBinding.getRoot())方法前定义。此时setContentView(R.layout.main)的命令代码行可以删除。

4.2　Button

Button(按钮)用来提示用户执行某种操作,属于TextView的子类。Button除了具有TextView的特性以外,很重要的一个特性就是响应用户单击操作的交互性。Button又派生出ImageButton、ToggleButton、RadioButton、CheckBox和Switch等控件。

4.2.1 单击监听器

监听器是 Android 控件响应用户操作的回调方法。对于 Button 控件使用最多的是单击监听器方法 OnClickListener()。代码如下:

视频讲解

【main.xml】
```xml
01  <?xml version = "1.0" encoding = "UTF - 8"?>
02  < LinearLayout xmlns:android = "http://schemas.android.com/apk/res/android"
03      android:layout_width = "match_parent"
04      android:layout_height = "match_parent"
05      android:orientation = "vertical">
06
07      < TextView
08          android:id = "@ + id/textView1"
09          android:layout_width = "match_parent"
10          android:layout_height = "50dp"
11          android:text = "@string/hello" />
12
13      < LinearLayout
14          android:id = "@ + id/linearLayout1"
15          android:layout_width = "match_parent"
16          android:layout_height = "wrap_content"
17          android:orientation = "horizontal"
18          android:weightSum = "1">
19
20          < Button
21              android:id = "@ + id/button1"
22              android:layout_width = "100dp"
23              android:layout_height = "wrap_content"
24              android:text = "按钮文字右对齐" />
25
26          < Button
27              android:id = "@ + id/button2"
28              android:layout_width = "100dp"
29              android:layout_height = "wrap_content"
30              android:text = "背景变灰右对齐" />
31      </LinearLayout >
32  </LinearLayout >
```

布局文件定义了一个垂直线性布局,其内包含一个 TextView 和一个垂直线性布局 linearLayout1,linearLayout1 内又包含两个 Button。按钮单击后的响应和处理在 Java 文件中实现。下面的代码要实现单击 button1 后,button1 内的文字右对齐;单击 button2 后,linearLayout1 内的控件(button1 和 button2)右对齐。

【FirstActivity.java】
```java
01  public class FirstActivity extends Activity
02  {
03      @Override
```

```
04      public void onCreate(Bundle savedInstanceState)
05      {
06          super.onCreate(savedInstanceState);
07          setContentView(R.layout.main);
08
09          Button button1;
10          Button button2;
11          button1 = (Button) findViewById(R.id.button1);
12          button2 = (Button) findViewById(R.id.button2);
13          final TextView textView1 = (TextView) findViewById(R.id.textView1);
14          final int i = 0;
15          button1.setOnClickListener(new View.OnClickListener()
16          {
17              @Override
18              public void onClick(View v)
19              {
20                  button1.setGravity(Gravity.RIGHT);
21                  textView1.setText("i = " + i);
22              }
23          });
24
25          button2.setOnClickListener(new View.OnClickListener()
26          {
27              @Override
28              public void onClick(View v)
29              {
30                  textView1.setBackgroundColor(Color.GRAY);
31                  textView1.setTextColor(Color.YELLOW);
32                  textView1.setText("背景已经设置为灰色!");
33                  LinearLayout linearLayout = (LinearLayout) findViewById(R.id.linearLayout1);
34                  linearLayout.setGravity(Gravity.RIGHT);
35              }
36          );
37      }
38  }
```

第15~23行定义了button1的setOnClickListener()设置单击监听器方法,本质上是1行代码,setOnClickListener()方法的参数是一个使用"new View.OnClickListener()"的匿名类实现接口的。OnClickListener是一个接口,onClick()是其中的抽象方法。通过匿名类实现接口,必须重写onClick()方法实现。需要单击button1实现的相关代码就放在onClick()方法中。注意,第23行最后的分号是第15行代码的结尾。第20行调用button1的setGravity()方法将button1中的文字右对齐。

第25~36行定义了button2的单击监听器方法,其中第30行将textView1背景色设为灰色。第31行将textView1文字设为黄色。第33行获取布局文件中的线性布局。第34行将线性布局linearLayout1内的所有控件(含textView1、button1和button2)设为右对齐。

启动程序后，单击按钮前的状态如图 4-14 所示。

当分别单击两个按钮后会分别执行各自监听器中的 onClick() 方法，结果如图 4-15 所示。

图 4-14　单击按钮前的状态

图 4-15　单击两个按钮后的状态

需要注意的是，在 JDK 8 以前的版本中，在第 13 和 14 行必须要声明变量是 final。以 textView1 为例，在第 30～32 行要调用 textView1，而这几行属于内部匿名类引用本地变量，系统提示"错误：从内部类引用的本地变量必须是最终变量或实际上的最终变量"。textView1 是在 FirstActivity 类的 onCreate() 方法中定义的，属于本地变量。修改的方法有以下几种：

（1）将本地变量用 final 修饰。

（2）将本地变量变为成员变量，即放在 onCreate() 方法之外 FirstActivity 类之内的位置。

（3）将相关变量放到内部类中声明。

（4）将 JDK 升级到版本 8（含）以上。此时变量虽然不需要用 final 修饰，但属于事实上的 final 变量，不能修改值。

细心的读者会发现，同样为 final 变量，第 21 行如果改成 i++ 就出错，这是因为 int 变量 i 在第 14 行定义为 final，所以不能修改 i 的值。但第 32 行的 textView1 也是 final 修饰，但可以使用 setText() 方法变更值，这是因为 setText() 方法变更的是对象 textView1 中的属性值（类的成员变量值），但并未变更 textView1 对象的地址。

【提问】　如果第 21 行的 i 改成每次单击 button1 就进行加 1 操作，程序应该如何修改？

将第 14 行的 final 去掉，如果仅此而已，也只是将变量 i 变成"实际上的最终变量（不能变更其值）"，单击 button1 并不能变更 i 值。如果将第 21 行输出的 i 改成 ++i。修改后会出现错误提示，提示为"错误：从内部类引用的本地变量必须是最终变量或实际上的最终变量"。要完成上述提问的功能可采用以下方案：

（1）将去掉 final 关键字的第 14 行移到第 16 行之后，此时变量 i 成为内部匿名类的成员变量。

（2）将去掉 final 关键字的第 14 行移到第 5 行之后，此时变量 i 成为外部类的成员变量。

（3）将鼠标放置到红色波浪线标识的变量 i 上，按下弹出提示框中的快捷键 Alt+Shift+Enter，自动将 int 型变量 i 转换为 int 型数组。

对 final 修饰的数组或对象仍然可以修改其中的值(并没有改变数组或对象的地址)，但不能重新初始化。

4.2.2 监听器复用

观察上面的案例会发现两个按钮的单击监听器框架都是相同的，可以考虑将其复用，用以下代码替换之前的 Java 代码。

```java
01  View.OnClickListener onClickListener = new View.OnClickListener()
02  {
03      @Override
04      public void onClick(View v)
05      {
06          switch(v.getId())
07          {
08              case R.id.button1:
09                  textView1.setTextColor(Color.BLUE);
10                  textView1.setText(button1.getText());
11                  break;
12              case R.id.button2:
13                  textView1.setTextColor(Color.RED);
14                  textView1.setText(button2.getText());
15                  break;
16          }
17      }
18  };
19  button1.setText("文字变蓝色");
20  button2.setText("文字变红色");
21  button1.setOnClickListener(onClickListener);
22  button2.setOnClickListener(onClickListener);
```

第 1~18 行定义对象变量 onClickListener。其中，第 4 行重写 onClick()方法，方法的形参 v 代表被单击按钮的 View 类型变量(数据类型是 View，强制转换为 Button 后就是触发单击操作的按钮对象)。第 6 行用 v.getId()方法获取触发单击操作按钮的 id 值，可用来判断是哪一个按钮触发的单击监听器，之后转到相应 case 语句执行相应代码。

第 21~22 行将两个按钮绑定 onClickListener 单击监听器，由此实现单击监听器代码的复用，减少了代码量，提高了代码的可读性。

4.2.3 长按单击监听器

后续案例的布局文件如果没有特别设置将不再列出，读者可参看运行结果反推布局设置或查看案例库中对应案例代码。按钮的长按监听器在用户长按按钮时触发(按下约 0.5 秒以上，与是否抬离按钮无关)。

【FirstActivity.java】
```
01  public class FirstActivity extends Activity
02  {
03      @Override
04      public void onCreate(Bundle savedInstanceState)
05      {
06          super.onCreate(savedInstanceState);
07          setContentView(R.layout.main);
08
09          Button button1 = (Button) findViewById(R.id.button1);
10          TextView textView1 = (TextView) findViewById(R.id.textView1);
11
12          button1.setOnLongClickListener(new View.OnLongClickListener()
13          {
14              @Override
15              public boolean onLongClick(View v)
16              {
17                  textView1.append("\n触发了长按操作");
18                  return false;
19              }
20          });
21
22          button1.setOnClickListener(new View.OnClickListener()
23          {
24              @Override
25              public void onClick(View v)
26              {
27                  textView1.append("\n触发了单击");
28              }
29          });
30      }
31  }
32  }
```

第12～20行定义button1的设置长按单击监听器方法，结构基本与setOnClickListener类似。setOnLongClickListener中的onLongClick()方法需要返回布尔型数据。如果返回false，则执行完长按单击监听器后，继续执行本按钮触发的其他监听器；如果返回true，则不再执行本按钮的其他监听器。

第22～29行定义的button1的单击监听器，通过textView1的输出显示观察两种监听器的触发时机和先后顺序。setOnLongClickListener()运行结果如图4-16所示。

当持续按下按钮时触发长按单击监听器，从第17行开始执行代码。由于第18行返回false，因此当手指离开屏幕时触发单击监听器，执行第27行代码。如果快速单击按钮则只执行单击监听器。

图4-16　setOnLongClickListener()运行结果

如果将第18行的false改为true,则长按单击操作完成后不再触发单击监听器。

【注】 用户对按钮的潜意识想法就是单击,因此使用长按单击监听器需注意:

(1) 最好有文字提示用户使用长按操作。

(2) 可用于有培训或使用说明的App中。

(3) 长按单击监听器不但可用于按钮控件也可以用于其他View控件,如在TextView上长按弹出"复制""粘贴"菜单。

在第21行添加以下命令:

```
button1.setClickable(false);
```

该命令是将button1的单击监听器响应关闭(默认是true),但单击或长按离开都会触发单击监听器。将鼠标指针放在如图4-17所示的setOnClickListener()方法名上,单击弹出界面的右下角,弹出上下文菜单项。

图4-17 查看源码菜单项

选择Edit Source(或者上方工具栏中的 ✎ 图标,根据版本或设置不同,显示会有差异,在图4-17所示的上下文菜单中选择Show Toolbar菜单,显示带工具条的弹出框,如图4-18所示,多出一个工具条),会显示View.java文件。更快捷的方式是按Ctrl+鼠标左键单击setOnClickListener()方法名或者按快捷键F4,就会显示View.java文件。

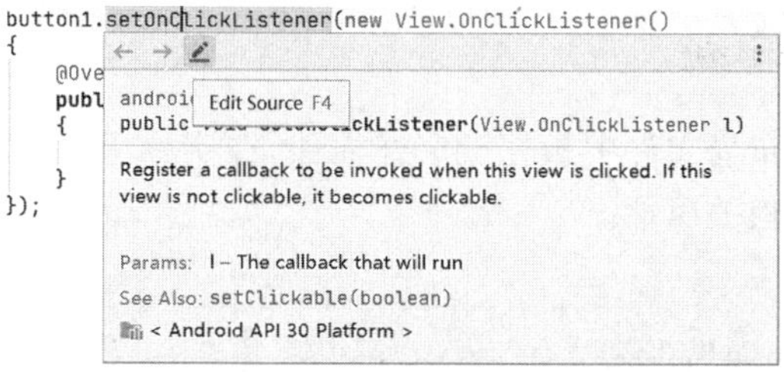

图4-18 带工具条的弹出框

```
public void setOnClickListener(@Nullable OnClickListener l) {
    if (!isClickable()) {
        setClickable(true);
    }
    getListenerInfo().mOnClickListener = l;
}
```

在 View.java 文件中找到 setOnClickListener()方法的定义,代码的含义是:控件的单击功能如果为 false 则设定为 true,即 setOnClickListener()方法会自动将控件的可单击属性设为 true。在本案例中可以使用"**button1.isClickable()**;"命令来获取 button1 是否可以单击。从上述代码可知,"**button1.setClickable(false)**;"必须设置在 setOnClickListener()之后才有禁用单击监听器效果。将此命令移动到第 30 行,重新运行程序,会发现还是会响应单击监听器。这应该算是缺陷。此时将第 12 行注释掉或者再加一条"**button1.setLongClickable(false)**;"将长按单击监听器的触发功能也禁用,再运行程序就能正常禁用 button1 的单击功能。

从人性化设计角度考虑,不建议使用"**button1.setClickable(false)**;"或"**button1.setLongClickable(false)**;"命令来禁用监听器,因为用户看到按钮就会认为按钮是可用的,但单击又没有效果,影响用户的使用体验。建议使用"**button1.setEnabled(false)**;"来替代,执行此命令会将按钮变为灰色,用户看到就知道此按钮当前处于不可用状态。也可以使用"**button1.setVisibility(View.INVISIBLE)**;"命令将按钮变为运行时不可见,或者用"**button1.setVisibility(View.GONE)**;"命令将按钮变为不可见且不占用显示空间。三者的显示效果如图 4-19～图 4-21 所示。

图 4-19 "button1.setEnabled(false);"
显示结果

图 4-20 "button1.setVisibility(View.INVISIBLE);"
显示结果

图 4-21 "button1.setVisibility(View.GONE);"显示结果

4.2.4 动态添加按钮

之前的案例都是在布局文件中预先设定好控件,在实际应用开发中可能会在程序运行时动态添加控件并设定相关属性。

```
【main.xml】
01  <?xml version = "1.0" encoding = "UTF - 8"?>
02  < LinearLayout xmlns:android = "http://schemas.android.com/apk/res/android"
03      android:layout_width = "match_parent"
04      android:layout_height = "fill_parent"
05      android:orientation = "vertical">
06
07      < LinearLayout
08          android:id = "@ + id/layoutButton"
09          android:layout_width = "match_parent"
10          android:layout_height = "match_parent"
11          android:orientation = "horizontal">
```

```
12
13          < Button
14              android:layout_width = "wrap_content"
15              android:layout_height = "wrap_content"
16              android:text = "我是布局文件中定义的按钮" />
17          <!-- 在此插入动态按钮 -->
18      </LinearLayout >
19 </LinearLayout >
```

布局文件结构是由一个垂直线性布局嵌套一个水平线性布局 layoutButton，在水平线性布局 layoutButton 内插入 9 个由 Java 代码生成的按钮，默认插入的按钮应该水平排列，但可以通过 Java 代码将 layoutButton 变为垂直方向布局。

【FirstActivity.java】
```
01 public class FirstActivity extends Activity
02 {
03      @Override
04      public void onCreate(Bundle savedInstanceState)
05      {
06          super.onCreate(savedInstanceState);
07          setContentView(R.layout.main);
08
09          LinearLayout.LayoutParams buttonParams;
10          final LinearLayout layout Button = (LinearLayout) findViewById
            (R.id.layoutButton);
11
12          layoutButton.setOrientation(LinearLayout.VERTICAL);
13          Button button;
14          for (int i = 1; i < 10; i++)
15          {
16              button = new Button(this);
17              button.setText("第" + i + "个动态生成按钮");
18
19              layoutButton.addView(button);  //将按钮添加到布局中
20              buttonParams = (LinearLayout.LayoutParams) button.getLayoutParams();
21              buttonParams.width = 300;
22              buttonParams.height = LinearLayout.LayoutParams.WRAP_CONTENT;
23              button.setLayoutParams(buttonParams);
24          }
25      }
26 }
```

第 9 行声明了一个获取信息布局相关参数的变量 buttonParams。

第 10 行获取布局文件中定义的水平线性布局 layoutButton。

第 12 行将 layoutButton 的水平线性布局改成垂直线性布局，此时布局文件中线性布局 layoutButton 的属性 android:orientation="horizontal" 被覆盖。

第 14~24 行使用 for 循环动态生成 9 个按钮。其中第 16 行动态创建一个按钮，第 17 行设置按钮的文字。此时生成的按钮对象还未绑定到布局中。第 19 行将动态创建的按钮通过 addView()方法绑定到线性布局 layoutButton 中。第 20 行获取刚刚加入布局中的按钮的相关布局参数。第 21~22 行设置按钮的宽度为 300px、高度以包含按钮文字为准。第 23 行将修改过的按钮参数重新同步到布局文件中的按钮属性中。动态生成按钮运行结果如图 4-22 所示。

因为高度问题有部分按钮无法显示，可以在代码中控制按钮的数量或高度来调整 9 个按钮的总体高度，或者使用 ScrollView 控件来实现滑动显示。

【注】 第 19 行的 addView()方法必须在第 20 行的 getLayoutParams()方法前，否则程序运行会出错。

图 4-22 动态生成按钮运行结果

4.2.5 自定义 DoubleClickListener 监听器

Android 的按钮有一系列的监听器，唯独没有双击监听器。本案例通过自定义的双击监听器类 DoubleClickListener 来实现类似双击的效果，读者也可以从中学习如何开发自定义功能的监听器。

【FirstActivity.java】
```
01   public class FirstActivity extends Activity
02   {
03       @Override
04       public void onCreate(Bundle savedInstanceState)
05       {
06           super.onCreate(savedInstanceState);
07           setContentView(R.layout.main);
08
09           final Button button1 = (Button) findViewById(R.id.button1);
10
11           button1.setOnClickListener(new DoubleClickListener()
12           {
13               @Override
14               public void onDoubleClick(View v)
15               {
16                   button1.setText("双击成功" + (new Date()).getTime());
17               }
18           });
19       }
20   }
21
22   abstract class DoubleClickListener implements View.OnClickListener
23   {
24       final long DOUBLE_TIME = 1000;
25       long lastClickTime = 0;
```

```
26
27        @Override
28        public void onClick(View v)
29        {
30            long currentTimeMillis = System.currentTimeMillis();
31            if (currentTimeMillis - lastClickTime < DOUBLE_TIME)
32            {
33                onDoubleClick(v);
34            }
35            lastClickTime = currentTimeMillis;
36        }
37        public abstract void onDoubleClick(View v);
38    }
```

此文件中除了 FirstActivity 类外,在第 22~38 行还定义了一个实现 OnClickListener 接口的 DoubleClickListener 抽象类。实现双击的思想是重写 OnClickListener 中的 onClick()方法,通过相邻两次单击的触发间隔来判断是否为双击操作。第 24 行定义了双击最大间隔为 1000ms。第 25 行变量 lastClickTime 用于记录上一次的单击时间。第 28~36 行重写 OnClickListener 中的 onClick()方法,其中第 31 行判断如果两次单击的间隔小于双击间隔时间 DOUBLE_TIME 就执行第 33 行的 onDoubleClick()方法。onDoubleClick() 抽象方法在第 37 行声明,抽象方法没有方法体,由实现接口的类重写抽象方法,本例中由第 14~17 行的 onDoubleClick()方法实现。

程序中分别使用(**new Date()). getTime()和 System. currentTimeMillis**()来获取当前时间,获取的时间是一样的。

此时程序存在一个小缺陷:当连续单击按钮且间隔都小于 DOUBLE_TIME 时会连续触发 onDoubleClick()方法。正常的连续双击应该是第 1 次和第 2 次连击属于双击,第 2 次和第 3 次不应该算双击。只要将双击后的 lastClickTime 重新置零即可实现完美的双击效果。重写的 onClick()代码如下:

```
@Override
    public void onClick(View v)
    {
        long currentTimeMillis = System.currentTimeMillis();
        if (currentTimeMillis - lastClickTime < DOUBLE_TIME)
        {
            onDoubleClick(v);
            lastClickTime = 0;
        } else
        {
            lastClickTime = currentTimeMillis;
        }
    }
```

【提问】 如果要实现 3 连击监听器该如何编写代码?

4.3 EditText

用户可以在 EditText（输入框）控件中输入文字并显示，EditText 常用于输入非限定内容的场景。EditText 是 Android 变动比较频繁的控件。早期 Android Studio 版本在 Palette 中就只有 1 个 EditText 类型。由于 EditText 使用非常广泛，为适应不同的场景，后期将 EditText 单列出 12 种类型的输入框。但无论如何变动，在 Code 视图中本质上还是 EditText 标签内的 inputType 属性不同而已。

视频讲解

4.3.1 设置和获取文本

```
【main.xml】
01    <?xml version = "1.0" encoding = "UTF-8"?>
02    <LinearLayout xmlns:android = "http://schemas.android.com/apk/res/android"
03        android:layout_width = "match_parent"
04        android:layout_height = "match_parent"
05        android:orientation = "vertical">
06
07        <EditText
08            android:id = "@+id/editText1"
09            android:layout_width = "match_parent"
10            android:layout_height = "wrap_content"
11            android:hint = "布局文件中的 hint"
12            android:inputType = "none"
13            android:text = "" />
14
15        <Button
16            android:id = "@+id/button1"
17            android:layout_width = "match_parent"
18            android:layout_height = "wrap_content"
19            android:text = "@string/button1" />
20
21        <TextView
22            android:id = "@+id/textView1"
23            android:layout_width = "match_parent"
24            android:layout_height = "wrap_content"
25            android:text = "@string/hello" />
26    </LinearLayout>
```

第 7~13 行定义了一个 EditText，其中第 11 行定义了 hint 属性，当输入框中没有文本时用灰色文字显示提示。EditText 中显示 hint，如图 4-23 所示。对应的 Java 命令为"**editText1. setHint**("布局文件中的 hint ");"。由于移动设备的屏幕显示区域有限，有时给 EditText 加 TextView 作为输入框功能描述标签会显得布局紧张，hint 提示功能可以更方便布局设计。当

图 4-23 EditText 中显示 hint

输入字符时此提示会自动消失，这也意味着如果 EditText 有初始值，就不会显示 hint。第 12 行定义 EditText 为普通输入框。

【FirstActivity.java】
```
01    public class FirstActivity extends Activity
02    {
03        @Override
04        public void onCreate(Bundle savedInstanceState)
05        {
06            super.onCreate(savedInstanceState);
07            setContentView(R.layout.main);
08
09            EditText editText1 = (EditText) findViewById(R.id.editText1);
10            Button button1 = (Button) findViewById(R.id.button1);
11            TextView textView1 = (TextView) findViewById(R.id.textView1);
12
13            button1.setOnClickListener(new View.OnClickListener()
14            {
15                @Override
16                public void onClick(View v)
17                {
18                    if(editText1.getText().toString().equals(""))
19                    {
20                        Date date = new Date();
21                        SimpleDateFormat formatter = new SimpleDateFormat("当前时
                            间：yyyy-MM-dd HH:mm:ss");
22                        editText1.setText(formatter.format(date));
23                    }
24                    textView1.append("\n" + editText1.getText());
25                }
26            });
27        }
28    }
```

在 button1 的单击监听器中实现当 editText1 中没有字符时在 editText1 和 textView1 显示当前时间，有字符时在 textView1 中显示 editText1 的字符。

第 24 行用 editText1.getText() 来获取 editText1 中的字符文本，此时返回的是 android.text.Editable 类型，为了执行""\n" + editText1.getText()"操作，android.text.Editable 类型会隐式转换为 String 类型，所以在使用 getText() 方法输出字符串时不用强制转换。如果更进一步分析，setText() 和 append() 方法的参数都是 CharSequence 类型，CharSequence 是接口，String 是实现 CharSequence 接口的类。容易出错的是第 18 行，要将 editText1 中的文本通过 getText() 方法取出并判断是否为""，一定要加上 toString() 将 android.text.Editable 类型转换为 String 类型，然后进行 String 字符串的比较。如果直接使用 editText1.getText().equals("")，则调用的是 android.text.Editable 类型的 equals() 方法，实际调用的是 Object 的 equals() 方法，而 Object 的 equals() 方法使用的是恒等号（==）比较，所以 editText1.getText() 方法返回的 android.text.Editable 类型与 String 类型比较的

结果是 false。可在第 18 行设置断点查看值比较结果,如图 4-24 和图 4-25 所示。

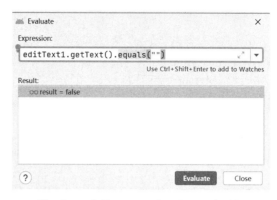

图 4-24　editText1.getText().equals("")

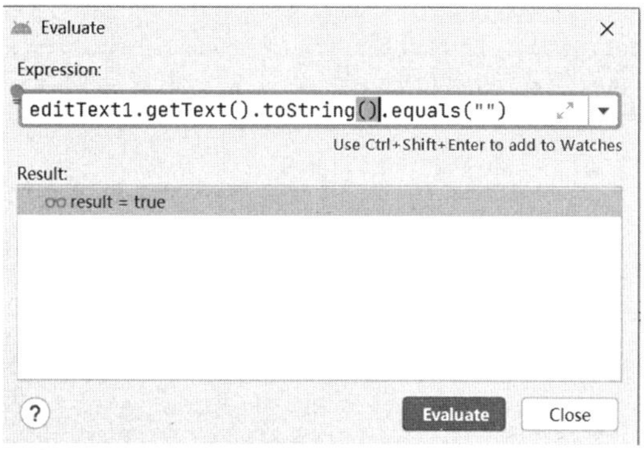

图 4-25　editText1.getText().toString().equals("")

如果将 onClick() 方法中的代码替换成以下代码,则 getText() 方法返回数据类型差异的运行结果如图 4-26 所示。

```
01  editText1.setText("OK");
02  textView1.append("\n editText1.getText().equals(\"OK\"): " + editText1.getText().equals("OK"));
03  textView1.append("\n editText1.getText().toString().equals(\"OK\"): " + editText1.getText().toString().equals("OK"));
04  textView1.append("\nbutton1.getText().equals(\"单击一下\"): " + button1.getText().equals("单击一下"));
```

图 4-26　getText() 方法返回数据类型差异的运行结果

同样是 getText()方法,使用 equals()方法会返回不同的结果,究其原因,输入框返回的是 Editable 类型,而按钮返回的是 CharSequence 类型。String 继承于接口 CharSequence,也可以是说 String 也是 CharSequence 类型。

4.3.2 按键监听器

当 EditText 获取焦点时,可以使用 setOnKeyListener 设置按键监听器来获取设备的硬按键按压信息,如按压音量键、返回键。本案例在布局文件中放置了两个 EditText,上面的 EditText 没有绑定按键监听器,不会因按下相应硬按键而触发按键监听器。下面的 EditText 获取焦点可触发按键监听器。

【FirstActivity.java】
```
01  public class FirstActivity extends Activity
02  {
03      @Override
04      public void onCreate(Bundle savedInstanceState)
05      {
06          super.onCreate(savedInstanceState);
07          setContentView(R.layout.main);
08
09          EditText editText1 = (EditText) findViewById(R.id.editText1);
10          editText1.setOnKeyListener(new View.OnKeyListener()
11          {
12              @Override
13              public boolean onKey(View v, int keyCode, KeyEvent event)
14              {
15                  editText1.setText("触发的 View: " + v.toString() + "\n 触发的 keyCode: " + keyCode + "\n 触发的 KeyEvent: " + event.toString());
16                  return true;
17              }
18          });
19      }
20  }
```

第 10~18 行在 editText1 上绑定按键监听器,其中第 13 行 onKey()方法的三个参数分别代表触发监听器的控件、音量键编码和事件(含按键按下或弹起状态、按键持续按下时间等)。第 16 行返回 true,如果按压音量键,将不会弹出音量调节状态条,也就不会变更系统的音量,可用于阅读小说软件的音量键翻页等场景。如果要继续执行音量键原有功能,可将返回值设为 false。运行程序时,如果 editText1(第二个输入框)获得焦点并且按下音量键,按键监听器运行结果如图 4-27 所示。如果持续按下相应音量键,repeatCount 和 eventTime 值会持续变化。按下和释放音量键,action 的值会依次显示 ACTION_DOWN 和 ACTION_UP。

图 4-27 按键监听器运行结果

4.3.3 触摸监听器

有些 App 如阅读软件在用户单击不同区域时选择前后翻页，要实现这样的功能可以使用触摸监听器。触摸监听器的触发事件分为按下（ACTION_DOWN）、移动（ACTION_MOVE）和弹起（ACTION_UP）。

```
【FirstActivity.java】
01   public class FirstActivity extends Activity
02   {
03       @Override
04       public void onCreate(Bundle savedInstanceState)
05       {
06           super.onCreate(savedInstanceState);
07           setContentView(R.layout.main);
08
09           EditText editText1 = (EditText) findViewById(R.id.editText1);
10           TextView textView1 = (TextView) findViewById(R.id.textView1);
11
12           textView1.setBackgroundColor(Color.LTGRAY); //设置背景色便于查看两控件的边界
13           textView1.setTextColor(Color.BLACK);
14
15           editText1.setOnTouchListener(new View.OnTouchListener()
16           {
17               @Override
18               public boolean onTouch(View v, MotionEvent event)
19               {
20                   editText1.setText("当前状态：" + event.getAction() + "\n坐标
                        X = " + event.getX() + ";坐标 Y = " + event.getY() + "\n事件
                        MotionEvent:\n" + event.toString());
21                   return false;
22               }
23           });
24
25           editText1.setOnClickListener(new View.OnClickListener()
26           {
27               @Override
28               public void onClick(View v)
29               {
30                   textView1.append("\n触发了单击监听器");
31               }
32           });
33       }
34   }
```

第 15～23 行定义了触摸监听器，其中第 20 行的 event.getAction() 方法返回 editText1 的触发事件是按下、移动或弹起。event.getX() 和 event.getY() 返回接触点的坐标，坐标原点为 editText1 的左上角。第 21 行返回 false，会在执行完触摸监听器后继续执行其他触发的监听器，如在 editText1 范围内离开屏幕，首先会触发触摸监听器的弹起事件，然后执行第 25 行定义的单击监听器事件。要强调的是，移动事件的触发是随着坐标的移动连续触发的,即使接触点已经离开 editText1 的范围(含 App 的标题栏和系统标题栏)。弹起时无论在什么位置都会触发触摸监听器,而单击监听器只在 editText1 范围内起效。

模拟器和真机上的运行结果是不同的,模拟器上按下并且不移动,只会触发 ACTION_DOWN,真机上会先触发 ACTION_DOWN,然后持续触发 ACTION_MOVE。触摸监听器运行结果如图 4-28 所示。

图 4-28　触摸监听器运行结果

4.3.4　焦点改变监听器

EditText 控件获取焦点时会自动弹出软键盘。早期版本的 Android 默认第一个输入框自动获取焦点,并自动调出软键盘。现在的 Android 默认任何输入框都未获得焦点。可以在布局文件或 Java 文件中设置默认获得焦点的输入框。

【main.xml】
```
01  <?xml version = "1.0" encoding = "UTF - 8"?>
02  < LinearLayout xmlns:android = "http://schemas.android.com/apk/res/android"
03      android:layout_width = "match_parent"
04      android:layout_height = "match_parent"
05      android:orientation = "vertical">
06
07      < EditText
08          android:id = "@ + id/editText1"
09          android:layout_width = "match_parent"
10          android:layout_height = "wrap_content"
11          android:inputType = "text" />
12
13      < EditText
14          android:id = "@ + id/editText2"
15          android:layout_width = "match_parent"
16          android:layout_height = "wrap_content"
17          android:inputType = "textPostalAddress"
```

```
18          android:selectAllOnFocus = "true"
19          android:text = "@string/edit2">
20           < requestFocus />
21      </EditText >
22
23      < Button
24          android:id = "@ + id/button1"
25          android:layout_width = "match_parent"
26          android:layout_height = "wrap_content"
27          android:text = "Focus(true)" />
28
29      < Button
30          android:id = "@ + id/button2"
31          android:layout_width = "match_parent"
32          android:layout_height = "wrap_content"
33          android:text = "InTouchMode(true)" />
34
35  </LinearLayout >
```

editText1 是用来测试获取或失去焦点的输入框，editText2 是用来获取焦点从而让 editText1 失去焦点的辅助测试输入框。

第 18 行的作用是当 editText2 获得焦点时自动将文本内容全选。

第 20 行在 EditText 标签之间，当布局文件显示在屏幕上时，editText2 自动获得焦点。程序第 1 次启动时获得焦点并全选文本，效果如图 4-29 所示。

图 4-29 获得焦点并全选文本

在布局文件中有两个按钮，分别实现 editText1 的禁用或使能获得焦点触发焦点改变监听器功能。

【FirstActivity.java】
```
01    public class FirstActivity extends Activity
02    {
03        @Override
04        public void onCreate(Bundle savedInstanceState)
05        {
06            super.onCreate(savedInstanceState);
07            setContentView(R.layout.main);
08
09            EditText editText1 = (EditText) findViewById(R.id.editText1);
10
11            editText1.setOnFocusChangeListener(new OnFocusChangeListener()
12            {
13                @Override
14                public void onFocusChange(View v, boolean hasFocus)
15                {
16                    if (hasFocus) //判断是否获得焦点
17                    {
18                        editText1.setText("当前输入框获得焦点!");
19                    } else
20                    {
21                        editText1.setText("当前输入框失去焦点!");
22                    }
23                }
24            });
25
26            Button button1 = (Button) findViewById(R.id.button1);
27            Button button2 = (Button) findViewById(R.id.button2);
28
29            button1.setOnClickListener(new OnClickListener()
30            {
31                @Override
32                public void onClick(View v)
33                {
34                    if (button1.getText().equals("Focus(true)"))
35                    {
36                        button1.setText("Focus(false)");
37                        editText1.setFocusable(false);//此时无法单击输入框获取焦点
38                        editText1.clearFocus();
39                    } else
40                    {
41                        button1.setText("Focus(true)");
42                        editText1.setFocusable(true);
43                        editText1.requestFocus();
44                    }
45                }
46            });
47
48            button2.setOnClickListener(new OnClickListener()
```

```
49              {
50
51                  @Override
52                  public void onClick(View v)
53                  {
54                      if (button2.getText().equals("InTouchMode(true)"))
55                      {
56                          button2.setText("InTouchMode(false)");
57                          editText1.setFocusableInTouchMode(false);//此时无法单击输
                                                                    //入框获取焦点
58                          editText1.clearFocus();
59                      } else
60                      {
61                          button2.setText("InTouchMode(true)");
62                          editText1.setFocusableInTouchMode(true);
63                          editText1.requestFocus();
64                      }
65                  }
66              });
67          }
68      }
```

第 11～24 行定义 editText1 的焦点改变监听器,其中第 14 行的布尔型变量 hasFocus 返回 editText1 是否获得焦点,true 为获取了焦点。

第 29～46 行 button1 的单击监听器用于设置 editText1 是否允许获取焦点。其中第 37 行设置 editText1 无法获取焦点。第 38 行清空 editText1 已获取的焦点(对于 setFocusable(false)命令,是否有此行效果都一样)。第 42 行设置 editText1 可以获取焦点。第 43 行将焦点落在 editText1 上。但实际运行程序时会发现第 1 次单击按钮时,editText1 确实失去了焦点,当再次单击此按钮时,editText1 却无法获取焦点。此种设置是否获取焦点的方式主要用在 TV 设备等有物理移动按键的场景。

button2 的作用与 button1 相似,只是将 setFocusable()方法换成了 setFocusableInTouchMode() 方法,主要用于触摸设备。button2 可随意控制 editText1 是否可获取焦点。

第 43 行的 requestFocus()也可以替换为 requestFocusFromTouch()方法,两种方法在这里的效果是完全一样的。但在某些情况下会略有不同。

requestFocus()方法在以下三种情况下无法获得焦点:

(1) 对应的 View 不支持 Focus。
(2) 对应的 View 支持 Focus,但是不支持在 Touch 模式下的 Focus,如 Button。
(3) 对应的 View 其祖先 View 设置了 FOCUS_BLOCK_DESCENDANTS 标志,阻止其子 View 获取焦点。

而 requestFocusFromTouch()方法设计的目的就是解决 requestFocus()在上述第二种不能获得焦点的情况下(Touch 模式下不支持焦点),也能够获得焦点使用的问题。

将第 43 行的代码换成"button1.requestFocusFromTouch();",当第二次单击 button2 按钮时会发现 button1 的颜色变深,是 button1 获得焦点的表现。

4.3.5 文本选择

对 EditText 中的文本进行复制操作的第 1 步是选择文本,常用的操作有用户手工选择或通过代码进行选择。上一个案例用代码选择所有文本,本案例使用 setSelection()方法选择指定的文本。很多科普资料中都提到一个汉字占 2 字节,真是这样吗?在回答这个问题之前要搞清楚文本长度和文本字节长度的区别。将字符串"Hi 安卓"保存到纯文本文件中,文件名使用编码格式名称,保存的编码格式分别为 GB 2312、GBK 和 UTF-8。文字占用字节数如图 4-30 所示。

图 4-30 文字占用字节数

"Hi 安卓"是两个字母两个汉字,gb2312.txt 占用空间是 6 字节,减去两个字母的 2 字节,剩余 4 字节表示两个汉字,所以 GB 2312 编码中每个汉字 2 字节。GB 2312 总共有 6763 个汉字。GBK 编码与 GB 2312 编码类似,区别在于 GBK 编码包含了更多的汉字,达到 21 866 个汉字。GBK 编码表示的汉字毕竟有限,还是有一些生僻字没有包含在 GBK 编码中,为此又推出了 GB 18030,包含 70244 个汉字。而 UTF-8 可视为国际化版本的编码格式,ASCII 字符只占 1 字节,希腊文、拉丁文等文字占 2 字节,常见汉字占 3 字节,生僻汉字占 4 字节。所以"Hi 安卓"作为常见汉字,每个汉字在 UTF-8 中占 3 字节,文件总共占 8 字节。而生僻字"懰"保存为"utf-8-4字节.txt",1 个汉字就占了 4 字节。Android 项目中的文件推荐使用 UTF-8 编码格式,它也是 Android Studio 默认的文件编码。如果使用 GBK 编码格式保存 Java 文件,在程序运行时可能会出现乱码,此时在 app 目录 build.gradle 文件的 android 标签内增加一行 compileOptions.encoding="GBK"就可以解决乱码问题。

```
【main.xml】
01   public class FirstActivity extends Activity
02   {
03       @Override
04       public void onCreate(Bundle savedInstanceState)
05       {
06           super.onCreate(savedInstanceState);
07           setContentView(R.layout.main);
08
09           TextView textView1 = (TextView)findViewById(R.id.textView1);
10           EditText editText1 = (EditText) findViewById(R.id.editText1);
11
12           editText1.setText("0123456789");                //设置显示内容
13           //editText1.setText("零壹贰叁肆伍陆柒捌玖");     //一个汉字一个字符
14
15           editText1.requestFocus();
16           editText1.setSelection(2, 7);
17
18           textView1.setText("文本长度: " + editText1.getText().length());
19           textView1.append("\n 文本字节长度: " + editText1.getText().toString().
                getBytes().length);
20       }
21   }
```

第 16 行 setSelection(int start，int stop)方法的第 1 个参数是起始字符位置索引，第 2 个参数是选择字符的终止字符位置索引，索引值从 0 开始。

第 16 行指明选择的字符从位置索引 2 到索引 7，即字符"23456"被选中。起始和终止索引对调效果是一样的，即第 16 行等效于"**editText1.setSelection(7，2);**"。

从 API 29 开始，需要增加第 15 行代码，editText1 获得焦点才能显示选中文本效果(早期的版本默认第 1 个输入框自动获得焦点，现在的版本默认所有 View 都未获得焦点)。ASCII 字符串的 setSelection()运行结果如图 4-31 所示。

第 18 行显示 editText1 中的字符数，正常情况下，ASCII 码和汉字都视为一个字符，但也有特例，如"㘚"会计文本长度为 2，字节长度为 4。

第 19 行是将 editText1 中的字符串用 getBytes()方法转换为字节码来计算字节数，是字符串真正的存储长度。

setSelection()方法是按字符而不是字节来选择的，因此将第 12 行的数字换成第 13 行的中文，中文字符串的 setSelection()运行结果如图 4-32 所示。

图 4-31 ASCII 字符串的 setSelection()运行结果　　图 4-32 中文字符串的 setSelection()运行结果

如果将第 16 行换成"**editText1.setSelection(7，12);**"，代码可以编译通过，但程序运行到此行时因字符串位置索引最大值为 9，而代码中要到 12，导致索引超出范围，程序直接弹出 App 错误提示，如图 4-33 提示。由于并非编译错误，Android Studio 的 build 窗口并不会显示错误提示。此时要查找故障代码可以通过设置断点并用 Debug 功能单步执行来查找出错位置。单步调试执行直至因错误的代码转到异常处理的位置为止就查到出错的代码行。对于此类可能导致异常的方法一定要加索引范围判断，如判断索引是否超出 editText1.getText().length()，或者将可能出问题的代码进行捕获异常处理。异常处理的方法参见后续章节。

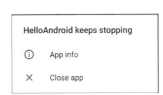

图 4-33 App 错误提示

4.3.6 禁止弹出软键盘

EditText 获得焦点后会自动弹出软键盘，在某些应用中希望禁用弹出软键盘功能，此时 EditText 非常类似 TextView，但 EditText 支持长按弹出粘贴提示，或者满足一定条件后恢复获得焦点功能(获得焦点也就会弹出软键盘实现字符输入)。在 4.3.7 节中还可以利用 inputType 属性实现比 TextView 更丰富的功能。

【FirstActivity.java】
```
01    public class FirstActivity extends Activity
02    {
03        @Override
04        public void onCreate(Bundle savedInstanceState)
```

```
05      {
06              super.onCreate(savedInstanceState);
07              setContentView(R.layout.main);
08
09              EditText editText1 = (EditText) findViewById(R.id.editText1);
10              //方法一:可以支持长按粘贴功能(如果没有定义 OnLongClickListener),
                //可以响应 OnClickListener
11          editText1.setFocusable(false);
12
13              //方法二:可以支持长按粘贴功能(如果没有定义 OnLongClickListener),
                //可以响应 OnClickListener
14              //editText1.setFocusableInTouchMode(false);
15
16              //方法三
17              //在布局文件中对 EditText 设置属性 android:focusable = "false"或者 android:
                //focusableInTouchMode = "false"
18
19              //方法四:在老版本 Activity 启动时并不弹出软键盘,当单击选取 EditText 控件
                //时才弹出软键盘.新版本 EditText 不弹出软键盘
20              //editText1.setKeyListener(null);
21
22              editText1.setOnClickListener(new View.OnClickListener()
23              {
24                  @Override
25                  public void onClick(View v)
26                  {
27                      editText1.append("Click");
28                  }
29              });
30
31              editText1.setOnLongClickListener(new View.OnLongClickListener()
32              {
33                  @Override
34                  public boolean onLongClick(View v)
35                  {
36                      editText1.append("LongClick");
37                      return false;
38                  }
39              });
40      }
41  }
```

代码中列出了 4 种禁止弹出软键盘的方法,如果定义了长按监听器,会影响系统自带弹出粘贴菜单的功能。方法一到方法三都通过禁止 editText1 获取焦点来防止软键盘弹出。方法四通过设定为 null 的 setKeyListener 监听器来防止软键盘弹出。

【提问】 如果禁用软键盘后又想输入文字该如何处理?
可采用下述任意方案。
(1) 满足条件时取消上述 4 种方法的对应代码,恢复 EditText 的文字输入功能。
(2) 使用弹出的对话框或调用其他 Activity 来返回输入文字。

4.3.7 inputType 和 imeOptions

视频讲解

Android Studio 一直在调整控件栏中 Text 类型控件的类型和数量,其本质就是拥有不同 inputType 和 imeOptions 属性的 EditText。无论如何调整,EditText 输入框所能展现的类型和外观都远远超过控件栏中列出的控件数量。inputType 属性可控制 EditText 获得焦点后弹出软键盘的类型,如 phone 类型 EditText 获得焦点后弹出手机拨号键盘,number 类型弹出纯数字键盘。合理配置 inputType 属性可优化用户使用体验。常用的 inputType 属性如下:

```
android:inputType = "text" 输入普通字符
android:inputType = "textCapCharacters" 字母大写
android:inputType = "textCapWords" 首字母大写
android:inputType = "textCapSentences" 仅第一个字母大写
android:inputType = "textAutoCorrect" 纠错
android:inputType = "textAutoComplete" 自动完成
android:inputType = "textMultiLine" 多行输入
android:inputType = "textImeMultiLine" 输入法多行(如果支持)
android:inputType = "textNoSuggestions" 不提示
android:inputType = "textUri" URI 格式,如网址
android:inputType = "textEmailAddress" 电子邮件地址
android:inputType = "textEmailSubject" 邮件主题
android:inputType = "textShortMessage" 短消息
android:inputType = "textLongMessage" 长消息
android:inputType = "textPersonName" 人名
android:inputType = "textPostalAddress" 地址
android:inputType = "textPassword" 密码
android:inputType = "textVisiblePassword" 可见密码
android:inputType = "textWebEditText" 作为网页表单的文本
android:inputType = "textFilter" 文本筛选过滤
android:inputType = "textPhonetic" 拼音输入
android:inputType = "number" 数字
android:inputType = "numberSigned" 带符号数字格式
android:inputType = "numberDecimal" 带小数点的浮点格式
android:inputType = "phone" 拨号键盘
android:inputType = "datetime" 时间日期
android:inputType = "date" 日期键盘
android:inputType = "time" 时间键盘
```

imeOptions 属性用于设定软键盘右下角功能键的外观和功能。具体显示的图标或文字会因中英文版或 Android 版本的不同而变化。

```
android:imeOptions = "actionNone" Enter 键,按下后光标到下一行
android:imeOptions = "actionSearch" 搜索
android:imeOptions = "actionSend" 发送
android:imeOptions = "actionPrevious" 上一项
android:imeOptions = "actionNext" 下一项
android:imeOptions = "actionGo" 前往,不隐藏软键盘
android:imeOptions = "actionDone" 前往,隐藏软键盘
```

inputType 和 imeOptions 对应常见软键盘参见图 4-34~图 4-39。

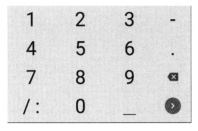

图 4-34　日期键盘

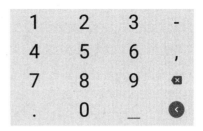

图 4-35　数字键盘

图 4-36　数字密码键盘

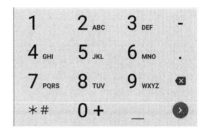

图 4-37　电话键盘

图 4-38　Email 键盘带发送键

图 4-39　文本键盘带查询键

【注】　对于 EditText 控件，要考虑不同的输入数据使用不同的软键盘以优化用户体验，还需要考虑优化布局，防止弹出的软键盘遮挡其他控件导致用户忽略被遮挡控件或焦点转入被遮挡控件困难的问题。

4.4　Toast

Toast 属于不能在布局文件中定义的视图，只能由 Java 代码定义并调用显示。当调用 Toast 时，Toast 的内容在屏幕上浮现一段时间后就会消失。调用 Toast 不会影响已呈现的布局。

4.4.1　显示文本

【FirstActivity.java】
```
01  public class FirstActivity extends Activity
02  {
```

```
03          @Override
04          public void onCreate(Bundle savedInstanceState)
05          {
06              super.onCreate(savedInstanceState);
07              setContentView(R.layout.main);
08
09              Toast msg1 = Toast.makeText(getApplicationContext(), "Toast 提示信息 1",
                    Toast.LENGTH_SHORT);
10              if (toast1.getDuration() == Toast.LENGTH_LONG)
11              {
12                  Log.i("xj", "Toast 提示信息 1 的持续时间为 LENGTH_LONG!");
13              } else
14              {
15                  Log.i("xj", "Toast 提示信息 1 的持续时间为 LENGTH_SHORT!");
16              }
17
18              toast1.show();
19              Toast.makeText(getApplicationContext(), "Toast 提示信息 2", Toast.LENGTH_
                    LONG).show();
20          }
21      }
```

第 9 行使用 makeText()方法生成显示文本的 Toast 对象,第 1 个参数是调用 Toast 的上下文 context,在此案例中可以用 this 指代,但推荐使用 getApplicationContext()方法以便适用于更多的场合,如封装在其他类中。第 2 个参数是要显示的文本。第 3 个参数是 Toast 显示持续时长,分为 Toast.LENGTH_SHORT(约 2s)和 Toast.LENGTH_LONG (约 3.5s)。由于显示时间有限,显示的文本内容应尽量精简。

第 10 行的 getDuration()方法用于获取显示持续时间类型,返回 int 型值,0 代表 Toast.LENGTH_SHORT,1 代表 Toast.LENGTH_LONG。

第 18 行用 show()方法将生成的 toast1 显示在屏幕上。

toast1 的使用遵循声明、定义、显示三个步骤。由于 Toast 具有显示后即消失的特性,而且很少重复使用,因此第 19 行将上述三个步骤合为一行的组合代码也被广泛使用。显示结果如图 4-40 所示。

图 4-40 Toast 显示文字

如果同时有多个 Toast 调用了 show()方法,Toast 会按照调用的顺序依次显示。长时间地遮挡正常布局显示且用户无法中断会给用户带来困扰。曾经有的版本修改为直接显示最后一个 Toast,这意味着有的信息无法显示。如果有重要信息提示用户,可在布局中添加显示反馈信息的控件,如 TextView,或者使用对话框。

4.4.2 显示图片

由于 Toast 显示时间较短,因此如果显示大量的文本会给用户带来困扰,此时显示图片能更好的向用户传递信息。

【FirstActivity.java】
```
01  public class FirstActivity extends Activity
02  {
```

```
03        @Override
04        public void onCreate(Bundle savedInstanceState)
05        {
06            super.onCreate(savedInstanceState);
07            setContentView(R.layout.main);
08
09            Toast toast1 = new Toast(getApplicationContext());
10            ImageView imageView1 = new ImageView(getApplicationContext());
11            imageView1.setImageResource(R.drawable.nophone1);
12            toast1.setView(imageView1);
13            toast1.show();
14        }
15    }
```

第 9 行先定义一个未填充任何信息的 toast1。

第 10 行使用 ImageView 构造方法定义一个对象 imagView1，同 Toast 构造方法一样可以使用 getApplicationContext()方法获取上下文 context 参数。

第 11 行将 imageView1 绑定到图片目录下的 nophone1.png 文件(Android 资源文件中的图片文件默认是 png 格式)。

第 12 行用 setView()方法将 imageView1 添加到 toast1 中。

第 13 行用 show()方法将 toast1 显示到屏幕中。

Toast 显示图片运行结果如图 4-41 所示。

案例中采用的是 100×100 像素的图片，如果改成大像素的广告图片覆盖屏幕，上述代码就变成显示几秒广告后就消失的程序。

图 4-41　Toast 显示图片运行结果

4.4.3　显示图片和文字

通过上述两个案例，可以使用 Toast 分别显示文字和图片。如果想同时显示图片和文字应该如何处理呢？对于 Toast，绑定文本的 makeText()方法和绑定图片的 setView()方法只能二选一，如果同时使用则只有最后一个执行的方法有效。要同时显示图文，需先使用 makeText()方法生成 Toast 实例，然后返回 Toast 实例中的布局，最后在布局中使用 addView()方法添加图片，具体代码如下：

【FirstActivity.java】
```
01    public class FirstActivity extends Activity
02    {
03        @Override
04        public void onCreate(Bundle savedInstanceState)
05        {
06            super.onCreate(savedInstanceState);
```

```
07          setContentView(R.layout.main);
08          Toast toast1 = Toast.makeText(getApplicationContext(), "带图片的Toast",
            Toast.LENGTH_LONG);
09          LinearLayout linearLayout1 = (LinearLayout) toast1.getView(); /
10          linearLayout1.setOrientation(LinearLayout.HORIZONTAL);
11          toast1.setGravity(Gravity.CENTER, 0, 0);
12          ImageView imageView1 = new ImageView(getApplicationContext());
13          imageView1.setImageResource(R.drawable.nophone1);
14          linearLayout1.addView(imageView1, 0);
15          toast1.show();
16      }
17  }
```

第8行使用makeText()方法将本文绑定到toast1。

第9行用getView()方法获取toast1中默认包含的线性布局并传给变量linearLayout1。

第10行将linearLayout1设定为水平线性布局(默认是垂直线性布局)。当前的toast1可视为一个线性布局中包含一个TextView控件，后续添加的图片会按水平方向排列在TextView控件后面。

第11行将toast1设置为屏幕水平和垂直方向都是中对齐(默认在屏幕下方)，对应的三个参数依次为：Toast在屏幕的基础定位、相对基础定位的X方向(水平方向)偏移量和相对基础定位的Y方向(垂直方向)偏移量。

第12～13行定义一个imageView并绑定图片文件nophone1.png。

第14行用addView()方法将imageView1添加到linearLayout1(linearLayout1之前已经包含使用makeText()方法添加的文本)中，此时linearLayout1中就同时包含文本和图片。

Toast显示图文运行结果如图4-42所示。

图4-42 Toast显示图文运行结果

此时toast1默认实现时长Toast.LENGTH_SHORT,可通过以下命令变更显示时长：

```
toast1.setDuration(Toast.LENGTH_LONG);
```

4.5 RadioButton

RadioButton(单选按钮)是常用基本控件之一。一般由多个单选按钮组合为一组实现从多个选项中选取一项。为了在多个单选按钮中形成排他性选择，引入RadioGroup控件来实现单选按钮的分组，同组内的单选按钮具有排他性，不同组的单选按钮没有排他性关联。

4.5.1 获取单选按钮选中项

【strings.xml】
```
01  <?xml version = "1.0" encoding = "UTF - 8"?>
02  < resources >
03      < string name = "hello">Hello World!你好,安卓!</string >
04      < string name = "app_name">HelloAndroid</string >
05      < string name = "textView1">请选择开发语言：</string >
06      < string name = "radioButton1">Java</string >
07      < string name = "radioButton2">C++</string >
08      < string name = "radioButton3">Python</string >
09      < string name = "radioButton4">C#</string >
10  </resources >
```

【main.xml】
```
01  <?xml version = "1.0" encoding = "UTF - 8"?>
02
03  < LinearLayout xmlns:android = "http://schemas.android.com/apk/res/android"
04      android:layout_width = "match_parent"
05      android:layout_height = "match_parent"
06      android:orientation = "vertical">
07
08      < TextView
09          android:id = "@ + id/textView1"
10          android:layout_width = "match_parent"
11          android:layout_height = "wrap_content"
12          android:text = "@string/textView1" />
13
14      < RadioGroup
15          android:id = "@ + id/radioGroup1"
16          android:layout_width = "match_parent"
17          android:layout_height = "wrap_content">
18
19          < RadioButton
20              android:id = "@ + id/radioButton1"
21              android:layout_width = "match_parent"
22              android:layout_height = "wrap_content"
23              android:text = "@string/radioButton1" />
24
25          < RadioButton
26              android:id = "@ + id/radioButton2"
27              android:layout_width = "match_parent"
28              android:layout_height = "wrap_content"
29              android:checked = "true"
30              android:text = "@string/radioButton2" />
31
32          < RadioButton
```

```
33              android:id = "@ + id/radioButton3"
34              android:layout_width = "match_parent"
35              android:layout_height = "wrap_content"
36              android:text = "@string/radioButton3" />
37
38          < RadioButton
39              android:id = "@ + id/radioButton4"
40              android:layout_width = "match_parent"
41              android:layout_height = "wrap_content"
42              android:text = "@string/radioButton4" />
43      </RadioGroup >
44
45  </LinearLayout >
```

第 14～43 行定义一个容器 RadioGroup,其内包含的 RadioButton 具有排他性,即同组内选中任意单选按钮都会导致其他单选按钮变为未选中状态。第 29 行指定 radioButton2 为选中状态。如果设为 false 或没有 android:checked 属性,则显示为未选中状态。

【FirstActivity.java】
```
01  public class FirstActivity extends Activity
02  {
03      @Override
04      public void onCreate(Bundle savedInstanceState)
05      {
06          super.onCreate(savedInstanceState);
07          setContentView(R.layout.main);
08
09          RadioGroup radioGroup1 = (RadioGroup) findViewById(R.id.radioGroup1);
10          RadioButton radioButton1 = (RadioButton) findViewById(R.id.radioButton1);
11          RadioButton radioButton2 = (RadioButton) findViewById(R.id.radioButton2);
12          RadioButton radioButton3 = (RadioButton) findViewById(R.id.radioButton3);
13          RadioButton radioButton4 = (RadioButton) findViewById(R.id.radioButton4);
14
15          radioGroup1.check(R.id.radioButton3);
16          radioGroup1.setOnCheckedChangeListener(new RadioGroup.OnCheckedChangeListener()
17          {
18              int count = 1;
19
20              @Override
21              public void onCheckedChanged(RadioGroup group, int checkedId)
22              {
23                  String str = "";
24
25                  //方法一:遍历,与是否使用监听器无关
26                  //if (checkedId == radioButton1.getId())
27                  //    str = (String) radioButton1.getText();
28                  //else if (checkedId == radioButton2.getId())
29                  //    str = (String) radioButton2.getText();
30                  //else if (checkedId == radioButton3.getId())
```

```
31              //     str = (String) radioButton3.getText();
32              //else if (checkedId == radioButton4.getId())
33              //     str = (String) radioButton4.getText();
34
35              //方法二：调用监听器的 checkedId 参数
36              //RadioButton radioButton = (RadioButton)findViewById(checkedId);
37              //str = (String)radioButton.getText();
38
39              //方法三：直接调用 RadioGroup 对象，与是否使用监听器无关
40              int radioButtonId = radioGroup1.getCheckedRadioButtonId();
41              RadioButton radioButton = (RadioButton)findViewById
                (radioButtonId);
42              str = radioButton.getText() + ",选中状态：" + radioButton.
                isChecked();
43
44              Log.i("xj", "你第" + count++ + "次选择的开发语言是：" + str);
45          }
46      });
47  }
48 }
```

第 16~46 行定义了 radioGroup1 的 OnCheckedChangeListener 监听器,当 radioGroup1 中的单选按钮选中状态有变化时(不单是选中,取消选中也算选中状态变化),触发此监听器并调用监听器中的 onCheckedChanged()方法,变量 checkedId 代表被选中单选按钮的 id。可使用下述 3 种方法获取选中单选按钮的文本。

第 26~33 行的方法一采用遍历 radioGroup1 中所有单选按钮的 id 是否等于 checkedId,如果相等就返回相应单选按钮的文本。此方法采用一系列的 if-else 来判定选中的单选按钮,代码稍显烦琐。可以将 if-else 换成 switch 优化代码可读性。

第 36~37 行的方法二是将 checkedId 直接用 findViewbyId()方法找到对应的单选按钮。此方法的优点是简单明了,可以无视 radioGroup1 中单选按钮的数量;缺点是要依靠 radioGroup1 的 OnCheckedChangeListener 监听器获取 checkedId。

第 40~42 行的方法三是对方法二的改进,通过 radioGroup1 的 getCheckedRadioButtonId() 方法获取选中的单选按钮 id。此方法可以不依赖 radioGroup1 监听器,推荐使用,特别适用于界面中同时有多个 RadioGroup 组成的单选按钮组,选定结果并提交时统一检查最终选定的单选按钮。

第 15 行是通过 radioGroup1 的 check()方法设置选中的单选按钮。需要注意,如果此行代码移动到 setOnCheckedChangeListener()方法之后,逻辑上变成 radioGroup1 先绑定 OnCheckedChangeListener 监听器,然后执行"radioGroup1.check(R.id.radioButton3);",将会触发监听器 3 次(具体可查看 RadioGroup 的 check()方法源码)：

(1) 如果已经有选中的单选按钮则设置为未选;
(2) 将当前选择的单选按钮设置为选中;
(3) 调用 OnCheckedChangeListener 监听器自己的方法。

程序运行的 Logcat 输出如下：

```
I:你第1次选择的开发语言是：C++,选中状态：false
I:你第2次选择的开发语言是：Python,选中状态：true
I:你第3次选择的开发语言是：Python,选中状态：true
```

因此为避免check()方法多次触发OnCheckedChangeListener监听器,应将check()方法放在监听器之前。或者将代码"radioGroup1.check（R.id.radioButton3）;"改成"radioButton3.setChecked(true);",直接对radioButton3执行setChecked()方法,此时将只触发radioGroup1的OnCheckedChangeListener监听器1次。

4.5.2 清空单选按钮

单选按钮一旦被选中,同一RadioGroup组中的所有单选按钮无法通过屏幕操作转为全部未选中状态,只能通过代码实现清空选择操作。本案例在4.5.1节案例的基础上增加一个按钮,在按钮单击监听器中执行清空单选按钮操作。

【FirstActivity.java】
```java
01  public class FirstActivity extends Activity
02  {
03      @Override
04      public void onCreate(Bundle savedInstanceState)
05      {
06          super.onCreate(savedInstanceState);
07          setContentView(R.layout.main);
08
09          RadioGroup radioGroup1 = (RadioGroup) findViewById(R.id.radioGroup1);
10          RadioButton radioButton1 = (RadioButton) findViewById(R.id.radioButton1);
11          RadioButton radioButton2 = (RadioButton) findViewById(R.id.radioButton2);
12          RadioButton radioButton3 = (RadioButton) findViewById(R.id.radioButton3);
13          RadioButton radioButton4 = (RadioButton) findViewById(R.id.radioButton4);
14          Button button1 = (Button) findViewById(R.id.button1);  //获取对象
15          Button button2 = (Button) findViewById(R.id.button2);
16          button1.setOnClickListener(new View.OnClickListener()
17          {
18              @Override
19              public void onClick(View v)
20              {
21                  //方法一
22                  radioGroup1.clearCheck();//清空选项,会触发OnCheckedChangeListener两
                                            //次,分别显示清空前后的id值,第二次为-1
23
24                  //方法二：触发OnCheckedChangeListener一次,显示清空前的值
25                  //int radioButtonId = radioGroup1.getCheckedRadioButtonId();
26                  //RadioButton radioButton = (RadioButton)findViewById
                    //(radioButtonId);
27                  //radioButton.setChecked(false);
28              }
29          });
30
```

```
31                //以下监听器用于演示清空效果
32                radioGroup1.setOnCheckedChangeListener(new RadioGroup.OnCheckedChangeListener()
33                {
34                        @Override
35                        public void onCheckedChanged(RadioGroup group, int checkedId)
36                        {
37                                String str = "";
38
39                                RadioButton radioButton = (RadioButton) findViewById(checkedId);
40                                try
41                                {
42                                        //radioButton 为 null 时,调用 radioButton.getText()会抛出异常
43                                        str = radioButton.getText() + "选中状态: " + radioButton.isChecked();
44                                } catch (Exception e)
45                                {
46                                        Log.i("xj", e.toString());
47                                }
48
49                                Log.i("xj", "id = " + checkedId + ",你选择的开发语言是: " + str);
50                        }
51                });
52
53                button2.setOnClickListener(new View.OnClickListener()
54                {
55                        @Override
56                        public void onClick(View v)
57                        {
58                                int radioButtonId = radioGroup1.getCheckedRadioButtonId();
59                                if (radioButtonId != -1)   //防止没有选择任何选项导致
                                                           //findViewById调用异常
60                                {
61                                        RadioButton radioButton = (RadioButton) findViewById(radioButtonId);
62                                        Toast.makeText(getApplicationContext(), "您选中了" +
                                                radioButton.getText(), Toast.LENGTH_LONG).show();
63                                }
64                        }
65                });
66        }
67 }
```

第 16～29 行 button1 的单击监听器中实现清空单选按钮的方法有如下两种。

方法一调用 radioGroup1 的 clearCheck()方法将同组内所有单选按钮都设置为未选。其优点是一条命令即可实现清空；缺点是如果设置了 radioGgroup1 的 OnCheckedChangeListener 监听器，clearCheck()方法将会触发两次 OnCheckedChangeListener 监听器，第一次将选中的单选按钮设置为未选，第二次触发 checkedId 为-1 的事件。值为-1 的 checkedId 执行第 39 行返回的 radioButton 值为 null，因此第 43 行对 null 值运行 getText()方法会抛出异

常。为保证程序继续执行,第 43 行要用 try-catch 捕获异常并在第 46 行输出异常原因。单击"清空"按钮时,程序运行结果如下:

```
01  I: id = 2131230978,你选择的开发语言是:C++选中状态:false
02  I: java.lang.NullPointerException: Attempt to invoke virtual method 'java.lang.CharSequence
    android.widget.RadioButton.getText()'on a null object reference
03  I: id = -1,你选择的开发语言是:
```

为避免触发两次监听器,可使用方法二。第 25 行通过 radioGroup1 的 getCheckedRadioButtonId()方法找到选中的单选按钮 id。第 26 行通过 id 找到对应的单选按钮。第 27 行将单选按钮通过 setChecked()方法直接设置为 false(注:此时 radioGroup1 的选中单选按钮 id 还是不变,可单击 button2 查看结果来验证,属于显示和属性值不同步的缺陷)。此时只触发 radioGgroup1 的 OnCheckedChangeListener 监听器一次。程序运行结果如下:

```
01  I: id = 2131230981,你选择的开发语言是:C++选中状态:false
```

使用上述两种方式清空单选按钮,再单击 button2,会出现两种不同的结果。对于方法一,第 59 行的 radioButtonId 为-1,不显示 Toast。对于方法二,第 59 行的 radioButtonId 为执行清空前的单选按钮 id 值,所以会执行第 62 行显示 Toast,并在 Toast 中显示清空前选中的单选按钮值。

【注】 一般很少用到清空单选按钮功能。如果用上述两种方法实现清空,一定要注意在下述两方面的局限性。

(1) OnCheckedChangeListener 监听器的触发次数问题。

(2) 单选按钮显示状态与 RadioGroup 的 getCheckedRadioButtonId()方法返回值的不同步问题。

4.6 CheckBox

CheckBox(复选框)可以单独使用,实现开关选择功能;也可以多个复选框组合使用,实现多重选择功能。复选框不需要类似 RadioGroup 的控件进行分组。复选框的属性和方法与单选按钮类似。

视频讲解

4.6.1 基本功能

复选框除了用户在 UI 界面进行手动选择以外还可以通过代码方式进行选中和反选操作,本案例讲解 3 种使用代码的方式来实现复选框反选操作。

```
【FirstActivity.java】
01  public class FirstActivity extends Activity
02  {
03      @Override
04      public void onCreate(Bundle savedInstanceState)
05      {
06          super.onCreate(savedInstanceState);
```

```
07          setContentView(R.layout.main);
08          CheckBox checkBox1 = (CheckBox) findViewById(R.id.checkBox1);
09          CheckBox checkBox2 = (CheckBox) findViewById(R.id.checkBox2);
10          CheckBox checkBox3 = (CheckBox) findViewById(R.id.checkBox3);
11          CheckBox checkBox4 = (CheckBox) findViewById(R.id.checkBox4);
12          Button button1 = (Button) findViewById(R.id.button1);
13          TextView textViewShow = (TextView) findViewById(R.id.textViewShow);
14          checkBox3.setChecked(true);  //当前复选框设为选中
15          button1.setOnClickListener(new View.OnClickListener()
16          {
17              @Override
18              public void onClick(View v)
19              {
20                  String str = "你选择的开发语言是：";
21                  //方法一
22                  if (checkBox1.isChecked())
23                  {
24                      checkBox1.setChecked(false);
25                  }
26                  else
27                  {
28                      checkBox1.setChecked(true);
29                      str += " " + checkBox1.getText();
30                  }
31                  //方法二
32                  checkBox2.setChecked(!checkBox2.isChecked());
33                  //方法三
34                  checkBox3.toggle();
35                  checkBox4.toggle();
36
37                  //遍历所有复选框,返回选中内容
38                  if (checkBox2.isChecked())
39                      str += " " + checkBox2.getText();
40                  if (checkBox3.isChecked())
41                      str += " " + checkBox3.getText();
42                  if (checkBox4.isChecked())
43                      str += " " + checkBox4.getText();
44                  textViewShow.setText(str);  //显示选择结果
45              }
46          });
47      }
48  }
```

第 14 行使用 checkBox3 的 setChecked(true)方法实现复选框选中,等效于布局文件中对 checkBox3 标签添加属性 android:checked="true"。

案例中提供了 3 种方法实现已选复选框的反选功能。

第 22～30 行的方法一是判断各复选框的选中状态,然后用 setChecked()方法实现反选。

第 32 行的方法二对当前复选框调用 isChecked()方法获取选中状态,调用 setChecked()方法设置当前状态取反后的值。

第 34 行方法三直接调用 toggle()方法实现反选,不用关心复选框原来是什么状态。

由于复选框不存在分组的概念,获取复选框的结果只能采用遍历或对各复选框使用监听器获取选中与否的状态。前者参见第 38～43 行,后者参见 4.6.2 节的案例。复选框反选运行结果如图 4-43 所示。

图 4-43 复选框反选运行结果

4.6.2 监听器

复选框继承于 CompoundButton,同时支持 OnClickListener 和 OnCheckedChangeListener 监听器。本案例使用监听器检测复选框状态的变化,同时明晰不同监听器的优先级问题。

```
【FirstActivity.java】
01  public class FirstActivity extends Activity
02  {
03      @Override
04      public void onCreate(Bundle savedInstanceState)
05      {
06          super.onCreate(savedInstanceState);
07          setContentView(R.layout.main);
08          CheckBox checkBox1 = (CheckBox) findViewById(R.id.checkBox1);
09          CheckBox checkBox2 = (CheckBox) findViewById(R.id.checkBox2);
10          CheckBox checkBox3 = (CheckBox) findViewById(R.id.checkBox3);
11          CheckBox checkBox4 = (CheckBox) findViewById(R.id.checkBox4);
12          TextView textViewShow = (TextView) findViewById(R.id.textViewShow);
13          //1. OnClickListener 监听器
14          checkBox1.setOnClickListener(new View.OnClickListener()
15          {
16              @Override
17              public void onClick(View v)
18              {
19                  if (checkBox1.isChecked())      //判断选中状态
20                  {
21                      //方法一
22                      textViewShow.append("\n" + checkBox1.getText() + "选中,来源: OnClickListener");
23                  } else
24                      //方法二: 可用于代码复用
25                      textViewShow.append("\n" + ((CheckBox) v).getText() + "取消,来源: OnClickListener");
26              }
27          });
28          //2. OnCheckedChangeListener 监听器,更常用
29          checkBox1.setOnCheckedChangeListener(new CompoundButton.OnCheckedChangeListener()
```

```
30      {
31          @Override
32          public void onCheckedChanged(CompoundButton buttonView, boolean isChecked)
33          {
34              if (isChecked)        //判断选中状态
35                  textViewShow.append("\n" + checkBox1.getText() + "选中,
                    来源:OnCheckedChangeListener");//输出复选框内容
36              else
37                  textViewShow.append("\n" + ((CheckBox)buttonView).getText() +
                    "取消,来源:OnCheckedChangeListener");
                    //可用于代码复用.可以直接简化为buttonView.getText()
38          }
39      });
40
41      checkBox2.setOnCheckedChangeListener(new CompoundButton.OnCheckedChangeListener()
42      {
43          @Override
44          public void onCheckedChanged(CompoundButton buttonView, boolean isChecked)
45          {
46              if (isChecked)
47                  textViewShow.append("\n" + checkBox2.getText() + "选中");
48              else
49                  textViewShow.append("\n" + checkBox2.getText() + "取消");
50          }
51      });
52      checkBox3.setOnCheckedChangeListener(new CompoundButton.OnCheckedChangeListener()
53      {
54          @Override
55          public void onCheckedChanged(CompoundButton buttonView, boolean isChecked)
56          {
57              if (isChecked)
58                  textViewShow.append("\n" + checkBox3.getText() + "选中");
59              else
60                  textViewShow.append("\n" + checkBox3.getText() + "取消");
61          }
62      });
63      checkBox4.setOnCheckedChangeListener(new CompoundButton.OnCheckedChangeListener()
64      {
65          @Override
66          public void onCheckedChanged(CompoundButton buttonView, boolean isChecked)
67          {
68              if (isChecked)
69                  textViewShow.append("\n" + checkBox4.getText() + "选中");
70              else
71                  textViewShow.append("\n" + checkBox4.getText() + "取消");
72          }
73      });
74  }
75 }
```

第 14～27 行使用 OnClickListener 监听器判断 checkBox1 的选中状态。在输出 checkBox1 的文本时如果没有复用监听器则推荐使用第 22 行的代码,反之推荐使用第 25 行的代码,其中的变量 v 代表触发单击监听器的复选框控件。

第 29～39 行使用 OnCheckedChangeListener 监听器判断 checkBox1 的选中状态变化。类似 OnClickListener 监听器,也可以使用 onCheckedChanged( ) 方法中的参数变量 buttonView 确定触发选中状态改变监听器的复选框。

运行程序时改变 checkBox1 的状态会先触发 OnCheckedChangeListener,然后再触发 OnClickListener。对于复选框控件推荐使用 OnCheckedChangeListener 监听器。

第 41 行开始的代码是结构类似 checkBox1 的 OnCheckedChangeListener 监听器,相同的代码尽量复用是开发人员遵循的一贯宗旨。实现监听器复用方案参见 4.6.3 节的案例。复选框监听器运行结果如图 4-44 所示。

图 4-44 复选框监听器运行结果

4.6.3 代码复用

【FirstActivity.java】
```
01  public class FirstActivity extends Activity
02  {
03      TextView textViewShow;
04
05      CompoundButton.OnCheckedChangeListener onCheckedChangeListener = new CompoundButton.OnCheckedChangeListener()
06      {
07          @Override
08          public void onCheckedChanged(CompoundButton buttonView, boolean isChecked)
09          {
10              if (isChecked)
11                  textViewShow.append("\n" + buttonView.getText() + "选中");
12              else
13                  textViewShow.append("\n" + buttonView.getText() + "取消");
14          }
15      };
16
17      @Override
18      public void onCreate(Bundle savedInstanceState)
19      {
20
21          super.onCreate(savedInstanceState);
22
23          setContentView(R.layout.main);
24
25          CheckBox checkBox1 = (CheckBox) findViewById(R.id.checkBox1);
26          CheckBox checkBox2 = (CheckBox) findViewById(R.id.checkBox2);
27          CheckBox checkBox3 = (CheckBox) findViewById(R.id.checkBox3);
```

```
28          CheckBox checkBox4 = (CheckBox) findViewById(R.id.checkBox4);
29          textViewShow = (TextView) findViewById(R.id.textViewShow);
30
31          //OnCheckedChangeListener 初始化代码也可以放在这里
32          //绑定监听器
33          checkBox1.setOnCheckedChangeListener(onCheckedChangeListener);
34          checkBox2.setOnCheckedChangeListener(onCheckedChangeListener);
35          checkBox3.setOnCheckedChangeListener(onCheckedChangeListener);
36          checkBox4.setOnCheckedChangeListener(onCheckedChangeListener);
37
38      }
39
40  }
```

上述代码的功能与 4.6.2 节案例相同，但代码量大大减少。

第 5～15 行定义了 OnCheckedChangeListener 监听器供 4 个复选框绑定监听器使用。其中第 8 行的变量 buttonView 对应选中状态发生变更的复选框控件。

第 33～36 行 4 个复选框分别调用 setOnCheckedChangeListener()方法绑定监听器。

【提问】 OnCheckedChangeListener 实例化代码可以放在第 39 行的位置吗？

当然可以。如果是在 FirstActivity 类中，可放置任意位置。如果是在 FirstActivity 类的方法中，只要在调用 onCheckedChangeListener 对象实例前的位置都可以。

4.7 CheckedTextView

视频讲解

CheckBox 的外观为选中框在左，文字紧靠选中框，比较适合问卷的多选择项场景。如果是软件的设置界面如，某选项是否激活的界面，往往是文字在左，选中框在屏幕右侧。此时可采用 TextView 加无文字的 CheckBox 组合完成类似布局。Android 提供了更简单的解决方案：CheckedTextView。CheckedTextView 不仅可用作复选框，还提供了更多的外观以满足不同的使用场合。

```
【main.xml】
01  <?xml version = "1.0" encoding = "UTF - 8"?>
02
03  < LinearLayout xmlns:android = "http://schemas.android.com/apk/res/android"
04      android:layout_width = "match_parent"
05      android:layout_height = "match_parent"
06      android:orientation = "vertical">
07
08      < CheckBox
09          android:id = "@ + id/checkBox1"
10          android:layout_width = "match_parent"
11          android:layout_height = "wrap_content"
12          android:text = "@string/checkBox1" />
13
14      < CheckedTextView
15          android:id = "@ + id/checkedTextView2"
```

```xml
16          android:layout_width = "match_parent"
17          android:layout_height = "wrap_content"
18          android:checkMark = "?android:attr/listChoiceIndicatorMultiple"
19          android:checked = "true"
20          android:text = "@string/checkedTextView2" />
21
22      <CheckedTextView
23          android:id = "@+id/checkedTextView3"
24          android:layout_width = "match_parent"
25          android:layout_height = "wrap_content"
26          android:checkMark = "?android:attr/listChoiceIndicatorMultiple"
27          android:text = "@string/checkedTextView3" />
28
29      <CheckedTextView
30          android:id = "@+id/checkedTextView4"
31          android:layout_width = "match_parent"
32          android:layout_height = "wrap_content"
33          android:checkMark = "?android:attr/listChoiceIndicatorSingle"
34          android:text = "@string/checkedTextView4" />
35
36      <CheckedTextView
37          android:id = "@+id/checkedTextView5"
38          android:layout_width = "match_parent"
39          android:layout_height = "wrap_content"
40          android:checkMark = "?android:attr/listChoiceIndicatorSingle"
41          android:text = "@string/checkedTextView5" />
42
43  </LinearLayout>
```

布局文件中放置了一个 CheckBox 和 4 个 CheckedTextView。

【注】 在 Design 视图中拖入 CheckedTextView 时一定要设置 android:checkMark 属性或在 Activity 中执行 setCheckMarkDrawable()方法，否则外观就是 TextView，无法显示选中框或相应图标。

【FirstActivity.java】
```java
01  public class FirstActivity extends Activity
02  {
03      @Override
04      public void onCreate(Bundle savedInstanceState)
05      {
06          super.onCreate(savedInstanceState);
07          setContentView(R.layout.main);
08          CheckBox checkBox1 = (CheckBox) findViewById(R.id.checkBox1);
09          CheckedTextView checkedTextView2 = (CheckedTextView) findViewById(R.id.checkedTextView2);
10          CheckedTextView checkedTextView3 = (CheckedTextView) findViewById(R.id.checkedTextView3);
11          CheckedTextView checkedTextView4 = (CheckedTextView) findViewById(R.id.checkedTextView4);
12          CheckedTextView checkedTextView5 = (CheckedTextView) findViewById(R.id.checkedTextView5);
```

```java
13
14          checkBox1.setOnCheckedChangeListener(new CompoundButton.OnCheckedChangeListener()
15          {
16              @Override
17              public void onCheckedChanged(CompoundButton buttonView, boolean isChecked)
18              {
19                  if (isChecked)
20                      Toast.makeText(getApplicationContext(), checkBox1.getText() +
                            "选中", Toast.LENGTH_SHORT).show();
21                  else
22                      Toast.makeText(getApplicationContext(), checkBox1.getText() +
                            "取消", Toast.LENGTH_SHORT).show();
23              }
24          });
25
26          //在布局文件中通过属性 android:checked = "true"设为选中
27          checkedTextView2.setOnClickListener(new View.OnClickListener()
28          {
29              @Override
30              public void onClick(View v)
31              {
32                  checkedTextView2.toggle();          //注意,需 toggle()方法反选状态,
                                                        //这点与 CheckBox 不同,本命令的
                                                        //位置须在判断选中状态之前
33                  if (checkedTextView2.isChecked())   //判断选中状态
34                  {
35                      Toast.makeText(getApplicationContext(), checkedTextView2.
                            getText() + "选中", Toast.LENGTH_SHORT).show();
36                  } else
37                  {
38                      Toast.makeText(getApplicationContext(), checkedTextView2.
                            getText() + "取消", Toast.LENGTH_SHORT).show();
39                  }
40
41              }
42          });
43
44          //R 前加 android,代表调用系统图片
45          checkedTextView3.setCheckMarkDrawable(android.R.drawable.star_off);
            //默认为选中,设置 star_off 图片,如果无此行则显示未选中方框
46          checkedTextView3.setOnClickListener(new View.OnClickListener()
47          {
48              @Override
49              public void onClick(View v)
50              {
51                  checkedTextView3.toggle();
52                  if (checkedTextView3.isChecked())
53                  {
54                      Toast.makeText(getApplicationContext(), checkedTextView3.
                            getText() + "选中", Toast.LENGTH_SHORT).show();
55
56                      checkedTextView3.setCheckMarkDrawable(android.R.drawable.
                            star_on);     //选中五角星,需匹配相应图片
```

```
57                 } else
58                 {
59                     Toast.makeText(getApplicationContext(), checkedTextView3.
                         getText() + "取消", Toast.LENGTH_SHORT).show();
60
61                     checkedTextView3.setCheckMarkDrawable(android.R.drawable.
                         star_off);           //未选中五角星
62                 }
63
64             }
65         });
66
67         checkedTextView4.setCheckMarkDrawable(android.R.drawable.arrow_down_float);
68         checkedTextView4.setOnClickListener(new View.OnClickListener()
69         {
70             @Override
71             public void onClick(View v)
72             {
73                 checkedTextView4.toggle();
74                 if (checkedTextView4.isChecked())
75                 {
76                     Toast.makeText(getApplicationContext(), checkedTextView4.
                         getText() + "选中", Toast.LENGTH_SHORT).show();
77                     checkedTextView4.setCheckMarkDrawable(android.R.drawable.
                         arrow_up_float);     //上箭头
78                 } else
79                 {
80                     Toast.makeText(getApplicationContext(), checkedTextView4.
                         getText() + "取消", Toast.LENGTH_SHORT).show();
81                     checkedTextView4.setCheckMarkDrawable(android.R.drawable.
                         arrow_down_float);   //下箭头
82                 }
83
84             }
85         });
86
87         //即使是唯一单选按钮也能通过单击变为未选状态,这是RadioButton不具有的特性
88         checkedTextView5.setOnClickListener(new View.OnClickListener()
89         {
90             @Override
91             public void onClick(View v)
92             {
93                 checkedTextView5.toggle();
94                 if (checkedTextView5.isChecked())
95                 {
96                     Toast.makeText(getApplicationContext(), checkedTextView5.
                         getText() + "选中", Toast.LENGTH_SHORT).show();
97                 } else
98                 {
99                     Toast.makeText(getApplicationContext(), checkedTextView5.
                         getText() + "取消", Toast.LENGTH_SHORT).show();
100                }
```

```
101
102                }
103            });
104        }
105 }
```

CheckedTextView 运行结果如图 4-45 所示。

从运行结果可看出 CheckedTextView 的使用更加灵活,但有几点需要注意:

(1) 可设置 checkMark 属性来模拟复选框、单选按钮或者其他图形。其中,单选按钮模式可以通过单击自身实现切换选中和未选状态,这是 RadioButton 不具备的(要取消选中状态只能通过命令来设置)。

(2) 单击切换 CheckedTextView 的选中状态外观并不意味着其 isChecked 状态也同步变动,还必须调用 toggle()方法来切换选中状态,这是与 RadioButton 的最大区别。

(3) 如第 45 行 CheckedTextView 的外观使用自定义图片,图片名称前缀为 android.R.drawable,是调用 Android 系统自带图标库。

图 4-45 CheckedTextView 运行结果

4.8 ImageView

视频讲解

ImageView 用于显示图片资源。本案例在布局文件中添加了 3 个 ImageView,除了 imageView1 使用 android:src 属性绑定了 icon.png 外,其他两个都未绑定图片资源,相关图片的绑定在 Java 文件中实现。

```
【FirstActivity.java】
01  public class FirstActivity extends Activity
02  {
03      @Override
04      public void onCreate(Bundle savedInstanceState)
05      {
06          super.onCreate(savedInstanceState);
07          setContentView(R.layout.main);
08          ImageView imageView1 = (ImageView) findViewById(R.id.imageView1);
09          ImageView imageView2 = (ImageView) findViewById(R.id.imageView2);
10          ImageView imageView3 = (ImageView) findViewById(R.id.imageView3);
11          buttonShow = (Button) findViewById(R.id.buttonShow);
12          Button buttonZoomIn = (Button) findViewById(R.id.buttonZoomIn);
13
14          buttonShow.setOnClickListener(new View.OnClickListener()
15          {
16              @Override
```

```
17                  public void onClick(View v)
18                  {
19                      //方法一: setImageResource
20                      imageView1.setImageResource(R.drawable.b1);
21                      imageView1.getLayoutParams().height = 40;    //设置高度会自动
                                                                     //按比例设置宽度
22                      imageView1.setScaleType(ImageView.ScaleType.FIT_START);
23                      //imageView1.setScaleType(ImageView.ScaleType.FIT_XY);
                        //观察两种 ScaleType 放大后的不同
24
25                      //方法二: setImageBitmap
26                      imageView2.setImageBitmap(BitmapFactory.decodeResource(getResources(),
                        R.drawable.flower));
27                      imageView2.getLayoutParams().height = 600;   //设置高度会自动
                                                                     //按比例设置宽度
28
29                      //注: 如果提供了完整的各种分辨率下的图片的话,以上两种方法都
                        //应该不会有混乱
30
31                      //方法三: setImageDrawable
32                      imageView3.setImageDrawable(getResources().getDrawable(R.drawable.
                        flower, null));                              //设置显示图片
33                      imageView3.setImageAlpha(100);               //0 - 255,超过 255 按 255 计,
                                                                     //255 为不透明
34                  }
35              });
36
37              buttonZoomIn.setOnClickListener(new View.OnClickListener()
38              {
39                  @Override
40                  public void onClick(View v)
41                  {
42                      //默认 getLayoutParams()得到的宽度和高度都是 0
43                      //前面设置过 imageView1.getLayoutParams().height = 40;
                        //getLayoutParams()会取到设置值
44                      ViewGroup.LayoutParams params1 = imageView1.getLayoutParams();
45
46                      params1.height += 100;
47                      //params1.width += 100;
48                      if (params1.height > 1000)
49                          imageView1.setImageResource(R.drawable.b2);
50                      imageView1.setLayoutParams(params1);
51                  }
52              });
53          }
54  }
```

本案例演示了 3 种绑定图片资源的方法。方法一的优点是命令简单,方法二隐含调用方法三,方法三的效率更高一些。

第 21 行实际是两条命令合二为一,getLayoutParams()是获取 imageView1 的布局参数,height 是布局参数中的高度。虽然只设置了高度,imageView1 的高度和宽度默认均按

比例缩放。setScaleType()方法设定imageView1的缩放方式。单击buttonZoomIn(放大)按钮,实现imageView1的放大效果。

第22行通过setScaleType()方法将imageView1控件的缩放方式设置为ImageView.ScaleType.FIT_START,当imageView1达到原图尺寸时将不再变化(第47行注释取消时除外)。

如果去掉第23行的注释,则imageView1不按比例缩放,且随指定尺寸变化而缩放。

第33行的setImageAlpha()方法设定图片的透明度。早期版本为setAlpha()方法,已不推荐使用,但网络资源上的很多代码还是使用setAlpha()。

第37~52行定义buttonZoomIn按钮的单击监听器。其中,第44行获取imageView1的布局参数并赋值给对象变量params1。第46行修改params1的高度值。第48~49行判断imageView1高度超过1000时切换图片,视觉上造成气球爆炸的效果。第50行将params1对象变量通过setLayoutParams()方法重新绑定到imageView1控件,params1的尺寸参数也就传递给imageView1。

连续单击"放大图片"按钮运行结果如图4-46所示。

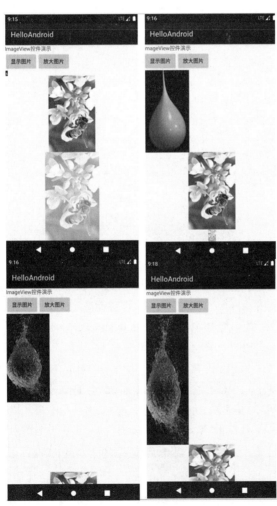

图4-46　连续单击"放大图片"按钮运行结果

4.9 DatePicker

DatePicker 是图形化获取日期的控件，可有效避免手工输入错误日期。DatePicker 的缺点是控件显示尺寸过大，造成布局设计美观度下降。因为还有一个类似的控件 CalendarView，所以在最新版本的 Android Studio 的控件栏中没有列出 DatePicker，但谷歌公司也没有将 DatePicker 列入弃用名单中。开发人员要使用 DatePicker，只能在 Code 视图中手工输入 DatePicker 标签。本书列举 DatePicker，是为了方便后续讲解有关联的、使用更加广泛的 DatePickerDialog。

```
【main.xml】
01  <?xml version = "1.0" encoding = "UTF - 8"?>
02
03  < LinearLayout xmlns:android = "http://schemas.android.com/apk/res/android"
04      android:layout_width = "match_parent"
05      android:layout_height = "match_parent"
06      android:orientation = "vertical">
07
08      < Button
09          android:id = "@ + id/button1"
10          android:layout_width = "match_parent"
11          android:layout_height = "wrap_content"
12          android:text = "获取日期" />
13
14      < DatePicker
15          android:id = "@ + id/datePicker1"
16          android:layout_width = "wrap_content"
17          android:layout_height = "wrap_content" />
18
19  </LinearLayout >
```

运行布局文件后 calendar 模式效果如图 4-47。单击年份数字可快速切换年份，这是 CalendarView 不具备的功能。

如果在 TimePicker 标签中增加属性 **android:datePickerMode=
"spinner"**，则 spinner 模式效果如图 4-48 所示。

如果在 AndroidManifest.xml 文件的 application 或 activity 标签(或者布局文件 TimePicker 标签)中增加属性 **android:theme="@android:style/Theme.Light"**，Theme.Light 风格效果如图 4-49 所示。此界面是早期 Android 版本 DatePicker 的默认外观。将 android:theme 属性添加到布局文件的 DatePicker 标签中也是同样的效果。

【注】 Android 会因为语言设置不同而导致年月日的显示顺序有差异。通过 color.xml 和 styles.xml 还可以修改颜色和主题等。

图 4-47 calendar 模式效果

图 4-48　spinner 模式效果

图 4-49　Theme.Light 风格效果

【FirstActivity.java】
```
01   public class FirstActivity extends Activity
02   {
03       @Override
04       public void onCreate(Bundle savedInstanceState)
05       {
06           super.onCreate(savedInstanceState);
07           setContentView(R.layout.main);
08
09           DatePicker datePicker1 = (DatePicker) findViewById(R.id.datePicker1);
10           Button button1 = (Button) findViewById(R.id.button1);
11
12           OnDateChangedListener onDateChangedListener = new OnDateChangedListener()
13           {
14               @Override
15               public void onDateChanged(DatePicker view, int year, int monthOfYear, int dayOfMonth)
16               {
17                   Toast.makeText(getApplicationContext(), "当前日期为: " + year +
                     "年" + (monthOfYear + 1) + "月" + dayOfMonth + "日", Toast.
                     LENGTH_LONG).show();                        //显示变更后的日期
18               }
19           };
20
21           //方法一：初始化日期并绑定日期变动监听器
22           datePicker1.init(2021, 1, 2, onDateChangedListener);
23
24           //方法二：更新当前日期
```

```
25              datePicker1.updateDate(2021, 1, 31);
26              if (Build.VERSION.SDK_INT >= Build.VERSION_CODES.O)  //单独绑定监听器,需
                                                                    //API 26 以上才支持
27              {
28                  datePicker1.setOnDateChangedListener(onDateChangedListener);
29              }
30
31              button1.setOnClickListener(new View.OnClickListener()
32              {
33                  @Override
34                  public void onClick(View v)
35                  {
36                      Toast.makeText(getApplicationContext(),"当前日期为: " + datePicker1.
                        getYear() + "年" + (datePicker1.getMonth() + 1) + "月" +
                        datePicker1.getDayOfMonth() + "日", Toast.LENGTH_LONG).show();
                        //显示日期信息
37                  }
38              });
39      }
40  }
```

第 12~19 行定义了 OnDateChangedListener 监听器,当选择的日期有变化时触发。实例化对象变量 onDateChangedListener 提供给 datePicker1 的 init()或 setOnDateChangedListener() 方法绑定监听器使用。

第 22 行的 init()方法可初始化日期并同时绑定 OnDateChangedListener 监听器。

第 25 行的 updateDate()方法用于更新 datePicker1 的选中日期,执行到此行会因改变了选中日期而触发 OnDateChangedListener 监听器。updateDate()方法也可以用于 datePicker1 的日期初始化,与 init()方法的区别是没有绑定监听器。早期 Android 版本绑定 OnDateChangedListener 监听器只能使用 init()方法,到了 API 26(Android 8.0/Oreo) 版本才开始支持 setOnDateChangedListener()方法来绑定 OnDateChangedListener 监听器,如第 28 行所示。为避免低版本设备执行此行代码,可采用两种方案来解决:

(1) 在 app 目录的 build.gradle 中将 minSdkVersion 设为 26 及以上版本。

(2) 添加第 26 行的代码,判断运行当前 App 的 Android 设备版本,条件成立才执行,否则忽略。

【注】 月份都是从 0 开始的,年和日都按正常年-数字计算。如果月和日超出正常范围,则折算成相应年数、月数或日数。

第 36 行使用 getYear()、getMonth()和 getDayOfMonth()分别获取年、月、日。不要忘记月份数加 1 才是实际月份数。

视频讲解

4.10 DatePickerDialog

DatePickerDialog 是日期对话框,不能在布局文件中设定,只能在 Activity 中通过 Java 代码调用。使用在类似如下的场景中:在需要输入日期的控件旁加一个按钮,单击按钮弹出日期对话框,选择相应日期后将选中日期返回到相应控件中显示。本案例的布局文件只有 button1 和 textView1,重点看 Java 文件。

【FirstActivity.java】
```
01  public class FirstActivity extends Activity
02  {
03      TextView textView1;
04
05      @Override
06      public void onCreate(Bundle savedInstanceState)
07      {
08          super.onCreate(savedInstanceState);
09          setContentView(R.layout.main);
10
11          textView1 = (TextView) findViewById(R.id.textView1);
12          Button button1 = (Button) findViewById(R.id.button1);
13
14          button1.setOnClickListener(new View.OnClickListener()
15          {
16              @Override
17              public void onClick(View v)
18              {
19                  //定义 DatePickerDialog 对象
20                  DatePickerDialog datePickerDialog1 = new DatePickerDialog(FirstActivity.
                        this, dateSetListener, 2021, 1, 3);
21                  datePickerDialog1.show();//显示 DatePickerDialog 对话框
22              }
23          });
24      }
25
26      DatePickerDialog.OnDateSetListener dateSetListener = new DatePickerDialog.
          OnDateSetListener()
27      {
28          @Override
29          public void onDateSet(DatePicker view, int year, int monthOfYear, int dayOfMonth)
30          {
31              textView1.setText("选择的日期：" + year + " - " + (monthOfYear + 1) +
                    " - " + dayOfMonth);
32          }
33      };
34  }
```

单击 button1,执行自第 20 行开始的代码,DatePickerDialog() 构造方法的第一个参数是 context。由于调用的 DatePickerDialog 日期对话框是依靠 View(本案例中的 button1) 调用的,其生命周期(相关内容参见后续章节)依赖于 FirstActivity,因此其构造方法的 context 是 FirstActivity 的 context。getApplicationContext() 方法获取的 context 是应用程序的 context,其生命周期与应用的生命周期相同,如果用于 DatePickerDialog() 构造方法的 context 会导致运行出错。第二个参数 dateSetListener 是在弹出的日期对话框中单击"确定"按钮后触发的监听器。最后三个参数是初始化选定日期的年、月(从 0 开始)、日。

需要注意 textView1 的声明和使用。由于第 31 行的 textView1 在内部类中调用,同时也要在 onCreate() 成员方法中调用,因此声明 textView1 是在第 3 行作为 FirstActivity 类的成员变量来声明的。DatePickerDialog 运行结果如图 4-50～图 4-52 所示。

图 4-50　DatePickerDialog　　图 4-51　弹出 DatePickerDialog　　图 4-52　关闭对话框返回日期
　　　　初始界面　　　　　　　　　　　　对话框　　　　　　　　　　　　　值 TimePicker

TimePicker 的功能与 DatePicker 类似，用于图形交互模式下获取时间。TimePicker 控件只能通过在布局文件中输入标签建立。

```
【main.xml】
01  <?xml version = "1.0" encoding = "UTF - 8"?>
02  < LinearLayout xmlns:android = "http://schemas.android.com/apk/res/android"
03      android:id = "@ + id/LL"
04      android:layout_width = "match_parent"
05      android:layout_height = "match_parent"
06      android:orientation = "vertical">
07
08      < TimePicker
09          android:id = "@ + id/timePicker1"
10          android:layout_width = "match_parent"
11          android:layout_height = "wrap_content" />
12
13      < Button
14          android:id = "@ + id/button1"
15          android:layout_width = "match_parent"
16          android:layout_height = "wrap_content"
17          android:text = "获取时间" />
18
19      < TextView
20          android:id = "@ + id/textView1"
21          android:layout_width = "match_parent"
22          android:layout_height = "wrap_content" />
23  </LinearLayout>
```

TimePicker 默认为 12 小时制的表盘外观，拖动数字上的圆形可调整小时，表盘数字随之变成分钟显示，再次拖动圆形调整分钟。可以单击"上午"或"下午"选择时段(12 小时制才有上午和下午的显示。英文版是在时间后面显示 AM 和 PM)，单击冒号前数字可调整小时，单击冒号后面数字可调整分钟。设置小时和分钟界面分别如图 4-53 和图 4-54 所示。

单击表盘左下角的图标 ▦ 转到软键盘设置时间界面,软键盘设置时间如图 4-55 所示,再次单击时间左下角的 ⏰ 图标返回表盘界面。

图 4-53　设置小时界面　　　图 4-54　设置分钟界面　　　图 4-55　软键盘设置时间

　　如果想变更 TimePicker 外观,可以在 AndroidManifest.xml 文件的 activity 标签(或者布局文件 TimePicker 标签)中添加属性 android:theme="@android:style/Theme.Light",Theme.Light 主题如图 4-56 所示,这是早期 Android 的 TimePicker 外观。单击"＋""－"符号修改时间,也可以单击数字后用软键盘修改时间。

　　将属性改为 android:theme="@android:style/Theme.Holo.Light",Theme.Holo.Light 主题如图 4-57 所示。可以拖动数字或单击上/下方灰色数字修改时间,也可以单击数字后用软键盘直接输入时间。

图 4-56　Theme.Light 主题　　　　　图 4-57　Theme.Holo.Light 主题

【FirstActivity.java】
```
01  public class FirstActivity extends Activity
02  {
03      @Override
04      public void onCreate(Bundle savedInstanceState)
05      {
06          super.onCreate(savedInstanceState);
07          setContentView(R.layout.main);
08
09          TextView textView1 = (TextView) findViewById(R.id.textView1);
10          TimePicker timePicker1 = (TimePicker) findViewById(R.id.timePicker1);
11          Button button1 = (Button) findViewById(R.id.button1);
12
13          timePicker1.setHour(18);        //设置小时
14          timePicker1.setMinute(19);      //设置分钟
15
16          timePicker1.setIs24HourView(true); //设置24小时制显示,默认是12小时制,
                                               //注意表盘的变化
17          timePicker1.setOnTimeChangedListener(new TimePicker.OnTimeChangedListener()
18          {
19              @Override
20              public void onTimeChanged(TimePicker view, int hourOfDay, int minute)
21              {
22                  //显示选择的时间,只要有时间变动就会触发
23                  textView1.setText("选择时间: " + hourOfDay + ":" + minute);
24              }
25          });
26
27          button1.setOnClickListener(new View.OnClickListener()
28          {
29              @Override
30              public void onClick(View v)
31              {
32                  textView1.setText("当前时间: " + timePicker1.getHour() + ":" +
                        timePicker1.getMinute());
33              }
34          });
35      }
36  }
```

第13～14行使用setHour()和setMinute()方法设置小时和分钟,其中小时是按24小时制设置。早期版本(API 23以前)使用setCurrentHour()和setCurrentMinute()方法设置小时和分钟,现在已经弃用。

第16行将timePicker1的外观按24小时制显示,即不显示"上午""下午",小时的数字为0～23(外圈为1～12,内圈为13～24,如果是12小时制就只有1～12)。

第17行设置timePicker1的OnTimeChangedListener监听器,只要时间有变动就触发。

第27行设置button1的OnClickListener监听器,用于在textView1中显示timePicker1的当前时间。

4.11 TimePickerDialog

视频讲解

调用 TimePickerDialog 会弹出时间选择对话框,在单击"确定"按钮后返回设定的时间。相关命令与 DatePickerDialog 相似。

【FirstActivity.java】
```
01   public class FirstActivity extends Activity
02   {
03       @Override
04       public void onCreate(Bundle savedInstanceState)
05       {
06           super.onCreate(savedInstanceState);
07           setContentView(R.layout.main);
08   
09           Button button = (Button) findViewById(R.id.button1);
10           button.setOnClickListener(new View.OnClickListener()
11           {
12               @Override
13               public void onClick(View v)
14               {
15                   Calendar calendar = Calendar.getInstance();//获取当前时间
16                   TimePickerDialog timePickerDialog1 = new TimePickerDialog(FirstActivity.this, timeSetListener1,
17                       calendar.get(Calendar.HOUR_OF_DAY),//小时
18                       calendar.get(Calendar.MINUTE),     //分钟
19                       false);                            //是否为24小时制
20                   timePickerDialog1.show();
21               }
22           });
23       }
24   
25       TimePickerDialog.OnTimeSetListener timeSetListener1 = new TimePickerDialog.OnTimeSetListener()
26       {
27           @Override
28           public void onTimeSet(TimePicker view, int hourOfDay, int minute)
29           {
30               TextView textView1 = (TextView) findViewById(R.id.textView1);
31               textView1.setText("选择的时间:" + hourOfDay + ":" + minute);
32           }
33       };
34   }
```

第15行用 Calendar 的 getInstance() 方法获取当前时间,Date 类中 getHours() 方法已弃用,推荐使用 Calendar。

第16行初始化 timePickerDialog1 对象,参数有监听器、小时、分钟和是否为24小时制。注意,方法中含有多个参数,如果参数在不同的行同样可以插入行注释,如第17行

所示。

第 20 行用 show() 方法将 timePickerDialog1 显示到屏幕。TimePickerDialog 与 TimePicker 一样可以在 AndroidManifest.xml 文件中通过调整 android:theme 的值来改变外观。

第 25～33 行定义 TimePickerDialog 的 OnTimeSetListener 监听器,当单击弹出的时间对话框中的"确定"按钮时触发。其中,onTimeSet() 方法中的形参 hourOfDay 和 minute 分别对应时间对话框中选定的小时和分钟。

时间对话框运行结果分别如图 4-58 和图 4-59 所示。

图 4-58　弹出时间对话框

图 4-59　返回时间对话框设定的时间

4.12　CalendarView

CalendarView 早期可以列出日历周数编号、控制是否可以被单击选择日期等功能,后来在功能上逐渐与 DatePicker 发生重叠。DatePicker 更偏重日期选择,具有年份的快速选择功能,支持 DatePickerDialog。CalendarView 更多偏重日历显示,从控件栏上列出的控件看,有 CalendarView 而没有 DatePicker(但也没有列入弃用列表中),谷歌公司可能更推荐 CalendarView。

```
【FirstActivity.java】
01    public class FirstActivity extends Activity
02    {
03        @Override
04        protected void onCreate(Bundle savedInstanceState)
05        {
06            super.onCreate(savedInstanceState);
```

```
07        setContentView(R.layout.main);
08
09        Date date;
10        TextView textView1 = (TextView) findViewById(R.id.textView1);
11        CalendarView calendarView1 = (CalendarView) findViewById(R.id.calendarView1);
12
13        //calendarView1.setDate(0);
14        calendarView1.setFirstDayOfWeek(2);
15        //calendarView1.setBackgroundColor(Color.YELLOW);
16
17        //以下两行命令在新版 SDK 中无效
18        //calendarView1.setEnabled(false);        //日历控件不可单击选择日期,可以用
                                                    //isEnabled()返回状态
19        //calendarView1.setShowWeekNumber(true); //显示第几周编号,新版本不再推荐使用
20
21        //方法一:使用 Calendar 设定时间
22        Calendar calendar = Calendar.getInstance();
23        calendar.add(Calendar.MONTH, -2);        //设置最小日期为当前日期前两个月
24        calendarView1.setMinDate(calendar.getTimeInMillis());//设置最小可选日期
25
26        //设置最大可选日期
27        //方法二:使用 new Date()设定时间
28        date = new Date();    //当前时间
29        int number = 30;      //将要加减的天数
30        long maxTime = date.getTime() + (long) number * 24 * 60 * 60 * 1000;
31        calendarView1.setMaxDate(maxTime);
32
33        //方法三:将指定字符串转日期和时间
34        SimpleDateFormat formatter = new SimpleDateFormat("yyyy-MM-dd HH:mm:ss");
35        try
36        {
37            date = formatter.parse("2010-1-20 11:08:13");
38            long maxTime2 = date.getTime();
39            //calendarView1.setMaxDate(maxTime2);
40        } catch (Exception e)
41        {
42            Log.i("xj", e.toString());
43        }
44
45        calendarView1.setOnDateChangeListener(new CalendarView.OnDateChangeListener()
46        {
47            @Override
48            public void onSelectedDayChange(CalendarView view, int year, int month,
              int dayOfMonth)
49            {
50                String nowCalendar = calendar.getTime().toString();
51                String nowDateTime = new Date().toString();
52
53                String date = year + "年" + (month + 1) + "月" + dayOfMonth +
                  "日\ncalendar.getTime(): " + nowCalendar + "\nnew Date()." +
                  "toString(): " + nowDateTime + "\nMin:" + calendarView1.
                  getMinDate() + "\nMax:" + calendarView1.getMaxDate();
```

```
54                    textView1.setText(date);
55                }
56            });
57        }
58    }
```

如果没有第 13 行，calendarView1 默认选择当前日期为选中日期。第 13 行如果去掉注释，calendarView1 的当前选中日期将被 setDate() 方法设置为 0，即"1970-01-01 00:00:00"，参数的数据类型为 long，如果要显示此日期之前的日期则输入负数即可。如果 setDate()[如果没有 setDate() 方法则默认为当天日期]与 setMinDate() 同时设置了日期，则以两者中的大值为选中日期。

第 14 行设置日历第一列显示星期几，默认参数为 1 即第一列为周日。数值小于 1 或大于 7 第一列都为周日。

第 15 行将 calendarView1 的背景色改为黄色。

第 18 行原本是将 calendarView1 设置为不可单击改变日期，现在版本中此命令无效（但也没有列入弃用名单）。

第 19 行在早期版本中用于在日历左侧显示日历的周数，新版本已弃用，且运行无效。

案例中使用了 3 种方法来设置日期。

方法一：第 22 行使用 Calendar 类进行日期设置。第 22 行使用 getInstance() 方法获取当前日期（含时间）。第 23 行调用 calendar 的 add() 方法可对年、月、日进行加减计算。第 24 行设置日历显示的最小可选日期，小于此日期部分将显示为灰色且不响应单击选取操作。

方法二：第 28 行使用 Date 类的构造方法获取当前日期（含时间），用第 30 行的 getTime() 方法获取当前时间的长整型数值，然后将要加减的天数转换成毫秒数，再用第 31 行的 setMaxDate() 方法设置日历最大可选日期。long 型时间还可以使用"System.currentTimeMillis()"方法获取。

方法三：第 34 行采用 SimpleDateFormat 类设定要转换的字符串日期格式，其中 M 是月份，m 是分钟。第 37 行按 formatter 指定的格式将字符串转日期类型，formatter.parse() 方法会抛出异常，所以需要将其放置在 try-catch 中。第 39 行如果去掉注释，最大可选日期小于最小可选日期，程序运行到此会发生错误。虽然第 39 行已经放在 try-catch 中，但程序仍然会因出错而终止运行，因为 try-catch 只能处理异常而不能处理错误。

第 45 行定义 calendarView1 的 OnDateChangeListener 监听器，与 DatePicker 的 OnDateChangeListener 监听器的结构和功能都相同。

CalendarView 运行结果如图 4-60 所示。

【注】 DatePicker 和 CalendarView 控件都不支持农历，如需要相关功能可引入第三方控件。

图 4-60　CalendarView 运行结果

4.13 SeekBar

SeekBar 控件支持用户拖曳操作设置 int 型数值,如音视频播放器的进度条,相比数字字符输入方式显得更快速、便捷。根据应用场合的不同,SeekBar 也衍生出不同的外观和操作模式。

【main.xml】
```xml
01 <?xml version = "1.0" encoding = "UTF - 8"?>
02 < LinearLayout xmlns:android = "http://schemas.android.com/apk/res/android"
03     xmlns:tools = "http://schemas.android.com/tools"
04     android:id = "@ + id/LL"
05     android:layout_width = "match_parent"
06     android:layout_height = "match_parent"
07     android:orientation = "vertical"
08     tools:ignore = "ExtraText">
09
10     < EditText
11         android:id = "@ + id/editText1"
12         android:layout_width = "match_parent"
13         android:layout_height = "wrap_content" />
14
15     < SeekBar
16         android:id = "@ + id/seekBar1"
17         android:layout_width = "match_parent"
18         android:layout_height = "wrap_content" />
19
20     < Button
21         android:id = "@ + id/button1"
22         android:layout_width = "match_parent"
23         android:layout_height = "wrap_content"
24         android:textAllCaps = "false"
25         android:text = "设定 SeekBar 值" />
26
27     < Button
28         android:id = "@ + id/button2"
29         android:layout_width = "match_parent"
30         android:layout_height = "wrap_content"
31         android:textAllCaps = "false"
32         android:text = "获取 SeekBar 值" />
33
34     < SeekBar
35         android:id = "@ + id/seekBar2"
36         android:layout_width = "match_parent"
37         android:layout_height = "wrap_content"
38         android:theme = "@android:style/Theme.Light" />
39
40     <!-- 最大值为 10,步进为 1,每次移动 10%,不是平滑移动 -->
41     < SeekBar
42         android:id = "@ + id/seekBar3"
43         android:layout_width = "match_parent"
```

```
44              android:layout_height = "wrap_content"
45              android:max = "10"
46              android:progress = "2" />
47
48         < SeekBar
49              android:id = "@ + id/seekBar4"
50              style = "@android:style/Widget.SeekBar"
51              android:layout_width = "match_parent"
52              android:layout_height = "wrap_content" />
53
54         < SeekBar
55              android:id = "@ + id/seekBar5"
56              android:layout_width = "match_parent"
57              android:layout_height = "wrap_content"
58              android:theme = "@android:style/Theme.Holo"/>
59
60    </LinearLayout >
```

布局文件中列出了5种不同外观和操作模式的SeekBar,外观主要由android:theme来控制。其中,seekBar1是API 26以后版本的默认SeekBar外观,除了滑动线条较细以外,控件两端都与屏幕有一定的间隔。滑动条显示高度较大的seekBar2适合用于分辨率较低的设备(早期的Android设备默认SeekBar外观)或适老化App界面设计中,控件与屏幕边缘对齐,在真机上滑动时无法获得最大值。

第45行的android:max属性设定SeekBar的最大值为10,此值设置越大,SeekBar的平滑移动效果越强。

第46行设置SeekBar的android:progress属性设为2,代表初始移动了2个步进单位,seekBar3最后的效果是SeekBar分成10份,每次移动1份,开始的默认值为2。

SeekBar控件外观及运行状态分别如图4-61~图4-63所示。

 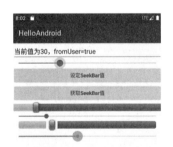 

图4-61　按下SeekBar状态　　　图4-62　移动SeekBar状态　　　图4-63　释放SeekBar状态

【FirstActivity.java】
```
01  public class FirstActivity extends Activity
02  {
03      @Override
04      public void onCreate(Bundle savedInstanceState)
05      {
06          super.onCreate(savedInstanceState);
07          setContentView(R.layout.main);
08          SeekBar seekBar1 = (SeekBar) findViewById(R.id.seekBar1);
09          EditText editText1 = (EditText) findViewById(R.id.editText1);
10          Button button1 = (Button) findViewById(R.id.button1);
11          Button button2 = (Button) findViewById(R.id.button2);
12  
13          seekBar1.setMax(100);
14  
15          seekBar1.setOnSeekBarChangeListener(new SeekBar.OnSeekBarChangeListener()
16          {
17              @Override
18              public void onStopTrackingTouch(SeekBar seekBar)
19              {
20                  editText1.setText("id 号为" + seekBar.getId() + "的 seekBar 停止更改");
21              }
22  
23              @Override
24              public void onStartTrackingTouch(SeekBar seekBar)
25              {
26                  editText1.setText("开始更改");
27              }
28  
29              @Override
30              public void onProgressChanged(SeekBar seekBar, int progress, boolean fromUser)
31              {
32                  editText1.setText("当前值为" + progress + ",fromUser = " + fromUser);
33              }
34          });
35  
36          button1.setOnClickListener(new View.OnClickListener()
37          {
38              @Override
39              public void onClick(View v)
40              {
41                  //String 转 int 方法一
42                  seekBar1.setProgress(Integer.valueOf(editText1.getText().toString()));
43                  //String 转 int 方法二
44                  //seekBar1.setProgress(Integer.parseInt(editText1.getText().toString()));
45              }
```

```
46              });
47
48              button2.setOnClickListener(new View.OnClickListener()
49              {
50                   @Override
51                   public void onClick(View v)
52                   {
53                        editText1.setText("获取值为" + seekBar1.getProgress());
54                   }
55              });
56         }
57   }
```

第 13 行用 setMax() 方法设置 seekBar1 的最大值,等效于布局文件中的属性 android:max。

第 15～34 行定义了 seekBar1 的 OnSeekBarChangeListener 监听器。当用户触摸 seekBar1 时触发 onStartTrackingTouch() 方法。滑动 seekBar1 时触发 onProgressChanged() 方法。手指离开屏幕时触发 onStopTrackingTouch() 方法。第 30 行的变量 progress 返回 seekBar1 的当前进度值。通过滑动 seekBar1 变更 progress 值时,变量 fromUser 返回 true,通过 Java 代码变更 progress 值时,变量 fromUser 返回 false。

第 36～46 行定义 button1 的单击监听器,将 editText1 中的数字通过 setProgress() 方法设定为 seekBar1 的进度值。第 42 行和第 44 行提供了两种 String 转 int 的方法,方法一的效率要略高于方法二。button1 的 onClick() 方法没有处理 editText1 中 String 转 int 是否会导致异常的问题,读者可自行将代码加以完善。

第 48～55 行定义 button2 的单击监听器,其中第 53 行的 getProgress() 方法可获取 seekBar1 的当前进度值。

视频讲解

4.14 RatingBar

RatingBar 通过选择星数量来替代分数输入,具有直观和便捷的特性。

```
【main.xml】
01   <?xml version = "1.0" encoding = "UTF - 8"?>
02   < LinearLayout xmlns:android = "http://schemas.android.com/apk/res/android"
03        android:id = "@ + id/LL"
04        android:layout_width = "match_parent"
05        android:layout_height = "match_parent"
06        android:orientation = "vertical">
07
08        < RatingBar
09             android:id = "@ + id/ratingBar1"
10             android:layout_width = "wrap_content"
11             android:layout_height = "wrap_content" />
12
13        < Button
14             android:id = "@ + id/button1"
```

```
15          android:layout_width = "match_parent"
16          android:layout_height = "wrap_content"
17          android:text = "@string/button1"
18          android:textAllCaps = "false" />
19
20      < Button
21          android:id = "@ + id/button2"
22          android:layout_width = "match_parent"
23          android:layout_height = "wrap_content"
24          android:text = "@string/button2"
25          android:textAllCaps = "false" />
26
27      < TextView
28          android:id = "@ + id/textView1"
29          android:layout_width = "match_parent"
30          android:layout_height = "wrap_content" />
31
32      <!-- 宽度设为 wrap_content,默认为五星, -->
33      <!-- 如果用 match_parent,排满一行,此时如果没有指定相关参数则与显示有差异 -->
34      < RatingBar
35          android:id = "@ + id/ratingBar2"
36          android:layout_width = "match_parent"
37          android:layout_height = "wrap_content">
38          <!-- 高度不能设为 match_parent,会导致显示异常 -->
39      </RatingBar>
40
41      <!-- ndroid:isIndicator = "true"只读 -->
42      < RatingBar
43          android:id = "@ + id/ratingBar3"
44          android:layout_width = "match_parent"
45          android:layout_height = "wrap_content"
46          android:isIndicator = "true" />
47
48      <!-- 中等、只读的 -->
49      < RatingBar
50          android:id = "@ + id/ratingBar4"
51          style = "@android:style/Widget.Holo.RatingBar.Indicator"
52          android:layout_width = "match_parent"
53          android:layout_height = "wrap_content" />
54
55      <!-- 小型、只读的 -->
56      < RatingBar
57          android:id = "@ + id/ratingBar5"
58          style = "@android:style/Widget.DeviceDefault.RatingBar.Small"
59          android:layout_width = "match_parent"
60          android:layout_height = "wrap_content" />
61
62      <!-- 按钮效果的 -->
63      < RatingBar
64          android:id = "@ + id/ratingBar6"
```

```
65              style = "@android:style/ButtonBar"
66              android:layout_width = "match_parent"
67              android:layout_height = "wrap_content" />
68      </LinearLayout>
```

ratingBar1 的宽度和高度都是 wrap_content，默认显示五星。

ratingBar2 在设置属性上与 ratingBar1 的区别在于控件宽度设为 match_parent，运行显示时会一直排列星号直至充满父容器。RatingBar 控件的高度如果设为 match_parent 将显示为空白。

ratingBar3 在第 46 行 android:isIndicator 属性为 true，指明 ratingBar3 是只读的，无法进行单击选择的交互操作。

ratingBar4、ratingBar5 和 ratingBar6 通过设置不同的 style 显示为不同的外观(带 Indicator 属性的是只读的)。

RatingBar 运行结果如图 4-64 所示。

图 4-64　RatingBar 运行结果

【FirstActivity.java】
```
01      public class FirstActivity extends Activity
02      {
03          @Override
04          public void onCreate(Bundle savedInstanceState)
05          {
06              super.onCreate(savedInstanceState);
07              setContentView(R.layout.main);
08              Button button1 = (Button) findViewById(R.id.button1);
09              Button button2 = (Button) findViewById(R.id.button2);
10              RatingBar ratingBar1 = (RatingBar) findViewById(R.id.ratingBar1);
11              TextView textView1 = (TextView) findViewById(R.id.textView1);
12
13              ratingBar1.setNumStars(5);              //设置星数量
14              ratingBar1.setStepSize((float) 0.5);    //设置步长,不建议使用小于 0.5
                                                        //的步长
15              ratingBar1.setRating(2.5f);             //设置当前分值
16
17              //ratingBar1.setIsIndicator(true);
18
19              ratingBar1.setOnRatingBarChangeListener(new RatingBar.OnRatingBarChangeLi
                stener()
20              {
21                  @Override
22                  public void onRatingChanged(RatingBar ratingBar, float rating, boolean
                    fromUser)
23                  {
24                      //显示相关变量值
```

```
25                        textView1.setText("当前值:" + rating + ",fromUser=" + fromUser);
26                    }
27                });
28
29                //setOnClickListener 无效
30                ratingBar1.setOnClickListener(new View.OnClickListener()
31                {
32                    @Override
33                    public void onClick(View view)
34                    {
35                        textView1.append("调用了 ratingBar1.setOnClickListener 方法");
36                    }
37                });
38
39                button1.setOnClickListener(new View.OnClickListener()
40                {
41                    @Override
42                    public void onClick(View v)
43                    {
44                        //显示 RatingBar 相关信息
45                        textView1.setText("当前值:" + ratingBar1.getRating() + "星,总
                            共:" + ratingBar1.getNumStars() + "星,步长:" + ratingBar1.
                            getStepSize() + ",只读:" + ratingBar1.isIndicator());
46                    }
47                });
48       }
49   }
```

第 13 行的 setNumStars() 方法可设置 ratingBar1 的星数量,默认 5 星。

第 14 行的 setStepSize() 方法设置步长值,默认为 0.5f,建议步长值为 0.5f 和 1.0f 两个值,选择其他数值虽然运行不会出错,但用户在单击时会出现显示与实际值不相符的问题(RatingBar 的宽度如果设置为 match_parent 也存在类似问题)。另外要注意浮点的赋值问题,以下列出了常见的浮点赋值。

```
float i = 0.5;          //会提示出错,默认小数视为 double 类型,double 类型无法隐式转换为 float
float a = (float) 0.5;  //正确,将 double 类型的 0.5 强制转换为 float
float b = 0.5f;         //正确,默认的 float 数据表示方式
```

第 15 行的 setRating() 方法设置当前分值,默认值为 0.0f。setStepSize 的设置值最好是 setRating 小数部分的倍数,为保证正常显示,setStepSize 的值最好只用 0.5f 和 1.0f,setRating 小数部分也为 0.5f 和 1.0f。如果上述两个方法的小数部分选择一致就按正常逻辑显示。如果 setStepSize 设置为 1.0f,setRating 的小数部分按四舍五入的方式显示选中的星号部分。

第 17 行的 setIsIndicator() 方法用于设置 ratingBar1 是否只读,如果设为 true,ratingBar1 就只能显示不能以交互方式选择。

第 19~27 行设置 OnRatingBarChangeListener 监听器,用于 ratingBar1 值变更时显示相关变量值。其中,变量 rating 用于记录 ratingBar1 变更后的值,变量 fromUser 记录变更

ratingBar1 值的来源,来自用户交互改变的为 true,来自 Java 代码变更的为 false。

第 30 行的 OnClickListener 监听器在 ratingBar1 中无效。

第 45 行展示 ratingBar1 相关方法获取控件属性值。

4.15 NumberPicker

视频讲解

NumberPicker 用于在设定的数值范围内选择 int 型数值。多个 NumberPicker 组合使用可实现如年、月、日、时、分、秒的自定义组合控件。

4.15.1 NumberPicker 基本功能

```xml
【main.xml】
01  <?xml version = "1.0" encoding = "UTF - 8"?>
02  < LinearLayout xmlns:android = "http://schemas.android.com/apk/res/android"
03      android:id = "@ + id/LL"
04      android:layout_width = "match_parent"
05      android:layout_height = "match_parent"
06      android:orientation = "vertical">
07
08      < NumberPicker
09          android:id = "@ + id/numberPicker1"
10          android:layout_width = "wrap_content"
11          android:layout_height = "80dp"/>
12      < Button
13          android:id = "@ + id/button1"
14          android:layout_width = "match_parent"
15          android:layout_height = "wrap_content"
16          android:text = "@string/button1"
17          android:textAllCaps = "false" />
18
19      < TextView
20          android:id = "@ + id/textView1"
21          android:layout_width = "match_parent"
22          android:layout_height = "wrap_content"
23          android:textColor = "@android:color/black"/>
24
25  </LinearLayout >
```

第 8~11 行定义 NumberPicker 控件,其中第 11 行将 numberPicker1 的高度设为 80dp,注意观察如果改成 wrap_content 后外观的变化。

```java
【FirstActivity.java】
01  public class FirstActivity extends Activity
02  {
03      @Override
04      protected void onCreate(Bundle savedInstanceState)
05      {
```

```java
06        super.onCreate(savedInstanceState);
07        setContentView(R.layout.main);
08
09        NumberPicker numberPicker1 = (NumberPicker) findViewById(R.id.numberPicker1);
10        Button button1 = (Button) findViewById(R.id.button1);
11        TextView textView1 = (TextView) findViewById(R.id.textView1);
12
13        numberPicker1.setTooltipText("请选择数值");//设置提示
14        numberPicker1.setMaxValue(30);              //设置最大值
15        numberPicker1.setMinValue(20);              //设置最小值
16        numberPicker1.setWrapSelectorWheel(true);   //设置数值是否头尾相接循环
17        numberPicker1.setValue(24);                 //设置当前值
18
19        numberPicker1.setOnValueChangedListener(new NumberPicker.OnValueChangeListener()
20        {
21            //当 NunberPicker 的值发生改变时,将会激发该方法
22            @Override
23            public void onValueChange(NumberPicker picker, int oldVal, int newVal)
24            {
25                textView1.setText("变更前:" + oldVal + ",变更后:" + newVal);
26            }
27        });
28
29        button1.setOnClickListener(new View.OnClickListener()
30        {
31            @Override
32            public void onClick(View v)
33            {
34                textView1.setText("最大值:" + numberPicker1.getMaxValue() + "\n最小值:" + numberPicker1.getMinValue() + "\n当前值:" + numberPicker1.getValue() + "\n提示信息:" + numberPicker1.getTooltipText() + "\n是否循环显示:" + numberPicker1.getWrapSelectorWheel());
35            }
36        });
37    }
38 }
```

第 13 行的 setTooltipText()方法设置控件提示,此提示需要长按控件才会弹出。

第 14 行的 setMaxValue()方法设置控件的最大值,默认值为 0,数值必须大于或等于 0,否则出错。

第 15 行的 setMinValue()方法设置控件的最小值,默认值为 0,数值必须大于或等于 0,否则出错。

第 16 行的 setWrapSelectorWheel()方法设置控件显示的 setMinValue 和 setMaxValue 数值是否头尾相接循环显示,默认值为 true。

第 17 行的 setValue()方法设置控件当前值。最终显示值由上面三行代码确定,具体如表 4-1 所示。

表 4-1 setValue()方法显示值

	条件 1	条件 2	显示值
1	setValue()＜setMinValue()	getWrapSelectorWheel()==false	getMinValue()
2	setValue()＜setMinValue()	getWrapSelectorWheel()==true	(getMaxValue()−getMinValue()) * x+setValue()+1
3	setValue()＞getMaxValue()	getWrapSelectorWheel()==false	getMaxValue()
4	setValue()＞getMaxValue()	getWrapSelectorWheel()==true	setValue()−(getMaxValue()−getMinValue()) * x−1
5	setMinValue()＜setValue()＜setMaxValue()		setValue()

【注】 Android 文档给出的显示值在表格第 2 行中为 getMaxValue()，第 4 行为 getMinValue()。但实际运行为表 4-1 中公式计算值。其中 x 的取值以显示值落在 [setMinValue(), setMaxValue()]区间为准。

第 19～27 行设置 OnValueChangeListener 监听器，当在图形界面滑动 numberPicker1 导致值发生变化时触发，同时可获取变更前后的数值。可以在 NumberPicker 控件上下滑动来改变数值，也可以长按上/下方的数字让数值连续变化来改变数值。

第 34 行输出 numberPicker1 控件的当前属性值。

NumberPicker 运行结果如图 4-65 和图 4-66 所示。

图 4-65 滑动 NumberPicker 控件

图 4-66 返回 NumberPicker 相关值

在实际应用中很少依靠单独的 NumberPicker 控件完成某项功能，如要设计一个同时可以设置年、月、日、时、分、秒的界面，需要 6 个 NumberPicker 加上说明的 TextView 共同来完成此项功能，所以也就没有所谓的 NumberPickerDialog。

4.15.2 显示文字的 NumberPicker

默认 NumberPicker 显示数字，可将字符串数组绑定到 NumberPicker 以实现文字的滚动选择效果。本案例的布局文件与 4.15.1 节的案例完全相同。

【FirstActivity.java】
```java
01  public class FirstActivity extends Activity
02  {
03    @Override
04    protected void onCreate(Bundle savedInstanceState)
05    {
06        super.onCreate(savedInstanceState);
07        setContentView(R.layout.main);
08
09        NumberPicker numberPicker1 = (NumberPicker) findViewById(R.id.numberPicker1);
10        Button button1 = (Button) findViewById(R.id.button1);
11        TextView textView1 = (TextView) findViewById(R.id.textView1);
12
13        String[] language = {"Java","C++","Python","C#","JavaScript","PHP"};
14        numberPicker1.setDisplayedValues(language);
15        numberPicker1.setMinValue(0);                           //设置显示的第一个数据
16        numberPicker1.setMaxValue(language.length - 1);         //设置显示的最后一个数据
17        numberPicker1.setTooltipText("请选开发语言");            //设置提示
18
19        //设置滑动监听器
20        numberPicker1.setOnValueChangedListener(new NumberPicker.OnValueChangeListener()
21        {
22            @Override
23            public void onValueChange(NumberPicker picker, int oldVal, int newVal)
24            {
25                textView1.setText("变更前:" + oldVal + ",变更后:" + newVal);
26                textView1.append("\n 变更前:" + language[oldVal] + ",变更后:" +
                         language[newVal]);
27            }
28        });
29
30        button1.setOnClickListener(new View.OnClickListener()
31        {
32            @Override
33            public void onClick(View v)
34            {
35                textView1.setText("最大值:" + numberPicker1.getMaxValue() + "\n 最
                         小值: " + numberPicker1.getMinValue() + "\n 当前值: " +
                         numberPicker1.getValue() + "\n 提示信息: " + numberPicker1.
                         getTooltipText() + "\n 是否循环显示: " + numberPicker1.
                         getWrapSelectorWheel());
36            }
37        });
38  }
39  }
```

代码与 4.15.1 节案例的代码基本相同,主要变化在第 13~17 行。

第 13 行定义了一个字符串数组 language。

第 14 行调用 numberPicker1 的 setDisplayedValues()方法将 language 数组绑定到

numberPicker1,此时 numberPicker1 不再显示数字而是显示字符串。

第 15 行 numberPicker1 的 setMinValue()方法设置显示的最小值,0 是数组索引最小值。

第 16 行 numberPicker1 的 setMaxValue()方法设置显示的最大值,也就是数组索引最大值,即数组长度减 1。

第 26 行通过参数 oldVal 和 newVal 作为字符串数组 language 的索引来显示变更前后的字符串值。

NumberPicker 运行结果如图 4-67 和图 4-68 所示。

图 4-67 滑动显示文字的 NumberPicker   图 4-68 返回显示文字的 NumberPicker 的相关值

视频讲解

4.16 ProgressBar

ProgressBar(进度条)控件有如下两种形态。

(1) style="?android:attr/progressBarStyle"的转圈的进度条。用于不确定时长的场景,早期版本叫不确定进度条,现在改为样式不同,而不确定模式变成了水平进度条的一个属性。

(2) style="?android:attr/progressBarStyleHorizontal"的水平进度条。

```
【main.xml】
01  <?xml version = "1.0" encoding = "UTF - 8"?>
02
03  < LinearLayout xmlns:android = "http://schemas.android.com/apk/res/android"
04      android:layout_width = "match_parent"
05      android:layout_height = "wrap_content"
06      android:orientation = "vertical">
07
```

```
08      <ProgressBar
09          android:id = "@ + id/progressBar1"
10          style = "?android:attr/progressBarStyleHorizontal"
11          android:layout_width = "300dip"
12          android:layout_height = "wrap_content"
13          android:max = "100"
14          android:progress = "70"
15          android:secondaryProgress = "80"
16          android:theme = "@android:style/Theme.Light" />
17
18      <TextView
19          android:layout_width = "wrap_content"
20          android:layout_height = "wrap_content"
21          android:text = "默认(主)进度条" />
22
23      <LinearLayout
24          android:layout_width = "match_parent"
25          android:layout_height = "wrap_content"
26          android:orientation = "horizontal">
27
28          <Button
29              android:id = "@ + id/button1"
30              android:layout_width = "wrap_content"
31              android:layout_height = "wrap_content"
32              android:text = "减少" />
33
34          <Button
35              android:id = "@ + id/button2"
36              android:layout_width = "wrap_content"
37              android:layout_height = "wrap_content"
38              android:text = "增加" />
39
40      </LinearLayout>
41
42      <TextView
43          android:layout_width = "wrap_content"
44          android:layout_height = "wrap_content"
45          android:text = "次进度条" />
46
47      <LinearLayout
48          android:layout_width = "match_parent"
49          android:layout_height = "wrap_content"
50          android:orientation = "horizontal">
51
52          <Button
53              android:id = "@ + id/button3"
54              android:layout_width = "wrap_content"
55              android:layout_height = "wrap_content"
56              android:text = "减少" />
57
```

```
58          < Button
59              android:id = "@ + id/button4"
60              android:layout_width = "wrap_content"
61              android:layout_height = "wrap_content"
62              android:text = "增加" />
63
64      </LinearLayout >
65
66      < ProgressBar
67          android:id = "@ + id/progressBar2"
68          style = "?android:attr/progressBarStyleHorizontal"
69          android:layout_width = "match_parent"
70          android:layout_height = "wrap_content"
71          android:max = "100"
72          android:progress = "70"
73          android:secondaryProgress = "80" />
74
75      < ProgressBar
76          android:id = "@ + id/progressBar3"
77          style = "?android:attr/progressBarStyleSmall"
78          android:layout_width = "match_parent"
79          android:layout_height = "wrap_content" />
80
81      < ProgressBar
82          android:id = "@ + id/progressBar4"
83          style = "?android:attr/progressBarStyle"
84          android:layout_width = "match_parent"
85          android:layout_height = "wrap_content" />
86
87      < ProgressBar
88          android:id = "@ + id/progressBar5"
89          style = "?android:attr/progressBarStyleLarge"
90          android:layout_width = "match_parent"
91          android:layout_height = "wrap_content" />
92
93      </LinearLayout >
```

布局文件中列出了常见的 ProgressBar 外观，根据改变 style 和 android:theme 的属性设置来变更外观。其中，android:theme＝"@android:style/Theme.Light"显示的就是早期 Android 版本的进度条外观。如果在 AndroidManifest.xml 中添加同样的属性，则布局中所有进度条都变为此主题的外观，与此同时，再添加以下代码可将标题栏也变成进度条：

```
requestWindowFeature(Window.FEATURE_PROGRESS);
```

不同风格、主题的 ProgressBar 如图 4-69 和图 4-70 所示。

第 13 行将 progressBar1 的进度条最大值设为 100，等效于 Java 命令"progressBar1.setMax(100)"。

第 14 行将 progressBar1 的进度条的主进度值设为 70。

图 4-69　不同风格、主题的 ProgressBar 之一

图 4-70　不同风格、主题的 ProgressBar 之二

第 15 行将 progressBar1 的进度条的次进度值设为 80。新老版本的水平进度条都能显示主、次进度，只是当前最新版本的默认进度条外观显示较细，主次进度条辨识度不高。

第 66～73 行定义当前默认水平进度条。

后续 3 个进度条都是不确定进度条，显示的大小有差异。

【FirstActivity.java】
```
01    public class FirstActivity extends Activity
02    {
03
04        @Override
05        protected void onCreate(Bundle savedInstanceState)
06        {
07            super.onCreate(savedInstanceState);
08
09            requestWindowFeature(Window.FEATURE_PROGRESS);    //显示标题栏进度条
10            setContentView(R.layout.main);
11
12            ProgressBar progressBar1 = (ProgressBar) findViewById(R.id.progressBar1);
13            ProgressBar progressBar2 = (ProgressBar) findViewById(R.id.progressBar2);
14            Button button1 = (Button) findViewById(R.id.button1);
15            Button button2 = (Button) findViewById(R.id.button2);
16            Button button3 = (Button) findViewById(R.id.button3);
17            Button button4 = (Button) findViewById(R.id.button4);
18
```

```
19          View.OnClickListener onClickListener = new View.OnClickListener()
20          {
21              @Override
22              public void onClick(View v)
23              {
24                  switch (v.getId())
25                  {
26                      case R.id.button1:
27                          progressBar1.incrementProgressBy(-1);//主进度减1
28                          setProgress(progressBar1.getProgress() * 100);
                            //标题栏进度条的范围为[0, 10000]
29                          break;
30                      case R.id.button2:
31                          progressBar1.incrementProgressBy(1);//主进度加1
32                          setProgress(progressBar1.getProgress() * 100);
33                          break;
34                      case R.id.button3:
35                          progressBar1.incrementSecondaryProgressBy(-1);
                            //次进度减1
36                          setSecondaryProgress(progressBar1.getSecondaryProgress()*100);
37                          progressBar2.setIndeterminate(true);
38                          break;
39                      case R.id.button4:
40                          progressBar1.incrementSecondaryProgressBy(1);
                            //次进度加1
41                          setSecondaryProgress(progressBar1.getSecondaryProgress()*100);
42                          progressBar2.setIndeterminate(false);
43                          break;
44                  }
45              }
46          };
47          button1.setOnClickListener(onClickListener);
48          button2.setOnClickListener(onClickListener);
49          button3.setOnClickListener(onClickListener);
50          button4.setOnClickListener(onClickListener);
51      }
52  }
```

单击各按钮分别实现主、次进度加减。第9行的作用是将标题栏也变成进度条。

第19～46行定义OnClickListener监听器实例onClickListener,供4个按钮绑定单击监听器实现代码复用。

第24行通过v.getId()方法获取触发监听器的按钮id,通过switch-case来执行不同按钮的单击响应代码。

第27行的incrementProgressBy()方法可实现progressBar1进度值的加减。

第28行的setProgress()方法实现对标题栏进度条的进度值加减。由于标题栏进度条的范围是[0, 10000],因此要将progressBar1通过getProgress()方法获取的进度值放大100倍以实现两种进度条的同步显示。setProgress()方法名有弃用标识(中画线)。谷歌在逐步弃用没有类前缀而是直接调用的方法。

第 37 行的"**progressBar2.setIndeterminate(true);**"是将 progressBar2 设置为不确定进度条,运行时注意观察改变前后运行状态的变化。

【注】 ProgressDialog 是带进度条的弹出对话框,由于进度条对话框影响用户与应用程序的交互,从 API 26 开始弃用。本书就不再讲解,感兴趣的读者可在案例库中自行查看相关代码。

4.17　Spinner

视频讲解

Spinner(下拉列表框)可提供用户在下拉列表中选择单一选项的功能,而且不能直接通过软键盘方式修改或增加选项,适用于对输入数据有一致性要求的场景。

```
【main.xml】
01  <?xml version = "1.0" encoding = "UTF - 8"?>
02  < LinearLayout xmlns:android = "http://schemas.android.com/apk/res/android"
03      android:layout_width = "match_parent"
04      android:layout_height = "match_parent"
05      android:orientation = "vertical">
06
07      < Spinner
08          android:id = "@ + id/spinner1"
09          android:layout_width = "match_parent"
10          android:layout_height = "wrap_content" />
11
12      < Spinner
13          android:id = "@ + id/spinner2"
14          android:layout_width = "match_parent"
15          android:layout_height = "wrap_content" />
16
17      < Spinner
18          android:id = "@ + id/spinner3"
19          android:layout_width = "match_parent"
20          android:layout_height = "wrap_content"
21          android:prompt = "@string/spinner3"
22          android:spinnerMode = "dialog" />
23
24      < Button
25          android:id = "@ + id/button1"
26          android:layout_width = "match_parent"
27          android:layout_height = "wrap_content"
28          android:text = "@string/button1" />
29
30      < TextView
31          android:id = "@ + id/textView1"
32          android:layout_width = "match_parent"
33          android:layout_height = "wrap_content" />
34
35  </LinearLayout>
```

在布局文件中放置了 3 个 Spinner,其中第 3 个 Spinner 在第 22 行增加了 android：spinnerMode="dialog"属性,单击此 Spinner 后其下拉列表将以对话框的形式显示。

【FirstActivity.java】

```
01   public class FirstActivity extends Activity
02   {
03       @Override
04       public void onCreate(Bundle savedInstanceState)
05       {
06           super.onCreate(savedInstanceState);
07           setContentView(R.layout.main);
08           Spinner spinner1 = (Spinner) findViewById(R.id.spinner1);
09           Spinner spinner2 = (Spinner) findViewById(R.id.spinner2);
10           Spinner spinner3 = (Spinner) findViewById(R.id.spinner3);
11           Button button1 = (Button) findViewById(R.id.button1);
12           TextView textView1 = (TextView) findViewById(R.id.textView1);
13
14           //方法一：动态初始化.使用 ArrayList,常用于从数据库循环遍历赋值
15           List<String> list1 = new ArrayList<String>();
16           list1.add("北京");
17           list1.add("上海");
18           list1.add("广州");
19           ArrayAdapter<String> arrayAdapter1 = new ArrayAdapter<String>(this, android.R.layout.simple_spinner_item, list1);    //初始化适配器
20           arrayAdapter1.setDropDownViewResource(android.R.layout.simple_spinner_dropdown_item);                //更改设置下拉列表的风格
21           spinner1.setAdapter(arrayAdapter1);             //绑定适配器
22           spinner1.setPrompt("请选择城市");              //设置标题
23           list1.add("深圳");                              //通过 list1 添加数据
24           //arrayAdapter1.add("新一线城市");              //通过适配器添加数据
25           //方法二：本地动态初始化.使用 String[]
26           String[] countries = {"中国", "美国", "英国", "法国", "俄罗斯"};
27           ArrayAdapter<String> arrayAdapter2 = new ArrayAdapter<String>(this, android.R.layout.simple_spinner_item, countries);    //初始化适配器
28           spinner2.setAdapter(arrayAdapter2);
29           spinner2.setPrompt("请选择五常国家");
30
31           //方法三：调用 strings.xml 中的数据进行初始化
32           ArrayAdapter<CharSequence> arrayAdapterPlanets = ArrayAdapter.createFromResource(this, R.array.planets, android.R.layout.select_dialog_item);
33           arrayAdapterPlanets.setDropDownViewResource(android.R.layout.simple_list_item_activated_1);                //显示选择的 item
34           //arrayAdapterPlanets.setDropDownViewResource(android.R.layout.select_dialog_singlechoice);        //更改设置单选下拉列表的对话框风格,为单选
35           //arrayAdapterPlanets.setDropDownViewResource(android.R.layout.select_dialog_multichoice);         //更改设置单选下拉列表的对话框风格,为多选
36           spinner3.setAdapter(arrayAdapterPlanets);
37           spinner3.setPrompt(getString(R.string.spinner3));
38
```

```
39          //方法一: spinner1 取值
40          spinner1.setOnItemSelectedListener(new AdapterView.OnItemSelectedListener()
41          {
42              @Override
43              public void onItemSelected(AdapterView <?> parent, View view, int position,
                long id)
44              {
45                  textView1.setText("选择的城市是: " + parent.getItemAtPosition
                    (position).toString() + "\nView 的值: " + ((TextView) view).
                    getText());
46              }
47
48              @Override
49              public void onNothingSelected(AdapterView <?> parent)//没有数据时触发
50              {
51                  textView1.append("\n 没有可选选项!");
52              }
53          });
54
55          //方法二: spinner1 取值
56          button1.setOnClickListener(new View.OnClickListener()
57          {
58              @Override
59              public void onClick(View v)
60              {
61                  textView1.setText("Spinner 标题为: " + spinner1.getPrompt() +
                    "\n 城市: " + spinner1.getItemAtPosition(spinner1.getSelected-
                    ItemPosition()) + "\n 城市: " + spinner1.getSelectedItem());
                    //获取标题和选中内容
62
63                  //list1.removeAll(list1);            //清空 list1 中数据
64                  //arrayAdapter1.notifyDataSetChanged();//通知适配器数据已变更,将触
                                                          //发 onNothingSelected()方法
65              }
66          });
67      }
68  }
```

开发人员并不能直接对 Spinner 的下拉选项的赋值,对下拉选项初始化过程类似于管道传送。首先定义数据源,其次定义连接数据源的适配器(管道),适配器的另外一端连接 Spinner。对数据源数据的变更通过适配器作用于 Spinner。本案例采用 3 种方法对 Spinner 进行赋值。

方法一: 第 15 行使用泛型建立 list1,第 16~18 行用 add()方法将字符串添加到 list1。此方式适用于将数据库中查询结果通过 add()方法进行循环赋值的场景。第 19 行初始化适配器 arrayAdapter1,其构造方法第一个参数是上下文,第二个参数是下拉列表的风格,第三个参数绑定数据源 list1。第 20 行的 setDropDownViewResource()方法用于变更下拉列表的风格。第 21 行将适配器绑定到 spinner1,此时数据源的数据就在 spinner1 上呈现。第 22 行的 setPrompt()方法将在下拉列表上显示标题提示。在新版本中 simple_spinner_dropdown_item 风格将不再显示标题提示。标题提示效果可参看 spinner3 中的显示。第

23~24行分别通过list1和arrayAdapter1来添加数据,变更后的数据都将呈现在spinner1的下拉列表中。但不建议使用第24行通过arrayAdapter1适配器添加数据,会导致OnItemSelectedListener监听器中的变量view返回null,当执行view的相关命令如((TextView)view).getText()时会因转换null而出错。

方法二：数据源采用字符串数组,适用于少批量固定数据录入。第26行定义字符串数组countries。第27行定义适配器arrayAdpter2,指定了下拉列表框风格并绑定countries数组。第28行将适配器绑定到spinner2。

```xml
【strings.xml】
<?xml version = "1.0" encoding = "UTF - 8"?>
< resources >
        < string name = "app_name">HelloAndroid </string>
        < string name = "button1">返回选中的值</string>
        < string - array name = "planets">
            < item >水星</item >
            < item >金星</item >
            < item >地球</item >
            < item >火星</item >
            < item >木星</item >
            < item >土星</item >
            < item >天王星</item >
            < item >海王星</item >
        </string - array >
        < string name = "spinner3">请选择星球</string >
</resources >
```

方法三：以strings.xml文件中的数据为数据源。第32行的R.array.planets对应strings.xml文件中的string-array name="planets"。第33~35行修改下拉列表框选中选项背景色、单选按钮风格和复选框风格(此时复选框也只能选择一项)。第37行的getString(R.string.spinner3)是从strings.xml文件中获取spinner3变量对应的值。

当用户通过spinner1下拉列表框选择选项后,获取返回值有以下两种办法。

方法一：使用spinner1的OnItemSelectedListener监听器获取值。当选择项有变化时触发onItemSelected()方法,当数据源清空并调用适配器的notifyDataSetChanged()方法时触发onNothingSelected()方法。第45行分别使用变量parent和变量view获取返回值。变量parent是指适配器,变量position返回下拉列表的选中项索引。表达式parent.getItemAtPosition(position).toString()是通过适配器的索引找到对应选项条目的文本。view是下拉列表的选中项,属于TextView类型,可使用((TextView)view).getText()返回选项文本。

方法二：通过spinner1的属性和方法获取选中的选项值。此方法一般用在用户提交时才获取用户最终的选择项。第61行先调用getSelectedItemPosition()方法获取spinner1的选中项索引,再调用getItemAtPosition()方法从索引值找到对应条目。更直接的方法是调用spinner1的getSelectedItem()方法获取选中条目。第63~64行演示触发OnItemSelectedListener监听器onNothingSelected()方法的条件。

不同风格的 Spinner 运行结果如图 4-71~图 4-73 所示。

图 4-71　simple_spinner_dropdown_item 风格的 Spinner

图 4-72　simple_spinner_item 风格的 Spinner

图 4-73　simple_list_item_activated_1 风格的 SpinnerToggleButton

4.18　ToggleButton

视频讲解

ToggleButton 是带状态的开关按钮，除了具有传统按钮的功能以外，外观和文字可以同时标识按钮的当前状态。

```
【main.xml】
01  <?xml version = "1.0" encoding = "UTF - 8"?>
02  < LinearLayout xmlns:android = "http://schemas.android.com/apk/res/android"
03    android:layout_width = "match_parent"
04    android:layout_height = "match_parent"
05    android:orientation = "vertical">
06
07    < ToggleButton
08        android:id = "@ + id/toggleButton1"
09        android:layout_width = "wrap_content"
10        android:layout_height = "wrap_content"
11        android:text = "ToggleButton" />
12
13    < ToggleButton
14        android:id = "@ + id/toggleButton2"
15        android:layout_width = "wrap_content"
16        android:layout_height = "wrap_content"
17        android:text = "ToggleButton"
18        android:theme = "@android:style/Theme.Light" />
19
```

```
20    <ToggleButton
21        android:id = "@ + id/toggleButton3"
22        android:layout_width = "wrap_content"
23        android:layout_height = "wrap_content"
24        android:text = "ToggleButton"
25        android:theme = "@android:style/Theme.Holo.Light" />
26
27    <ImageView
28        android:id = "@ + id/imageView1"
29        android:layout_width = "wrap_content"
30        android:layout_height = "wrap_content"
31        android:src = "@drawable/icon" />
32
33    <Button
34        android:id = "@ + id/button1"
35        android:layout_width = "match_parent"
36        android:layout_height = "wrap_content"
37        android:text = "测试 callOnClick 与 performClick 区别"
38        android:textAllCaps = "false" />
39
40    <Button
41        android:id = "@ + id/button2"
42        android:layout_width = "match_parent"
43        android:layout_height = "wrap_content"
44        android:text = "获取 toggleButton1 信息"
45        android:textAllCaps = "false" />
46
47 </LinearLayout>
```

布局文件中添加了 3 个 ToggleButton 控件，其中第 18 行和第 25 行指定的主题对应老版本 Android 控件外观。ToggleButton 运行结果如图 4-74 和图 4-75 所示。

图 4-74　ToggleButton 关闭状态　　　　图 4-75　ToggleButton 打开状态

【FirstActivity.java】
```
01  public class FirstActivity extends Activity
02  {
03      @Override
04      public void onCreate(Bundle savedInstanceState)
05      {
06          super.onCreate(savedInstanceState);
07          setContentView(R.layout.main);
08
09          ToggleButton toggleButton1 = (ToggleButton) findViewById(R.id.toggleButton1);
10          Button button1 = (Button) findViewById(R.id.button1);
11          Button button2 = (Button) findViewById(R.id.button2);
12          ImageView imageView1 = (ImageView) findViewById(R.id.imageView1);
13
14          imageView1.setVisibility(View.INVISIBLE);//隐藏图片,等效于 android:
                                                     //visibility="invisible"
15          toggleButton1.setChecked(true);          //设置按钮状态,默认为 false
16
17          toggleButton1.setTextOff("关灯");        //设置未选中的文本
18          toggleButton1.setTextOn("开灯");         //设置选中的文本
19
20          toggleButton1.setBackgroundDrawable(getResources().getDrawable(R.drawable.icon));
                                                     //设置按钮的背景
21          toggleButton1.setTextColor(Color.RED);   //改变按钮的文字颜色
22          toggleButton1.setOnCheckedChangeListener(new CompoundButton.OnCheckedChangeListener()
23          {
24              @Override
25              public void onCheckedChanged(CompoundButton buttonView, boolean isChecked)
26              {
27                  imageView1.setVisibility(View.VISIBLE);      //将图片设为可见
28                  //方法一
29                  if (isChecked)
30                  {
31                      imageView1.setImageResource(R.drawable.on);
32                  } else
33                  {
34                      imageView1.setImageResource(R.drawable.off);
35                  }
36                  //方法二:三元表达式.条件?a:b;
37                  imageView1.setImageResource(isChecked ? R.drawable.on : R.drawable.off);
38                  Log.i("xj", "触发 OnCheckedChangeListener");  //演示优先级
39              }
40          });
41
42          toggleButton1.setOnClickListener(new View.OnClickListener()
43          {
44              @Override
45              public void onClick(View view)
46              {
47                  Log.i("xj", "触发 OnClickListener");          //演示优先级
```

```
48              }
49          });
50
51          button1.setOnClickListener(new View.OnClickListener()
52          {
53              @Override
54              public void onClick(View view)
55              {
56                  toggleButton1.callOnClick();
57                  //toggleButton1.performClick();
58              }
59          });
60
61          button2.setOnClickListener(new View.OnClickListener()
62          {
63              @Override
64              public void onClick(View v)
65              {
66                  Toast.makeText(getApplicationContext(), "未选中文本：" + toggleButton1.
                    getTextOff() + ",选中文本：" + toggleButton1.getTextOn() + ",当前文
                    本：" + toggleButton1.getText(), Toast.LENGTH_LONG).show();
67              }
68          });
69      }
70  }
```

第14行用setVisibility()方法设置imageView1的可视状态，其值如下。

(1) VISIBLE：控件可见。

(2) INVISIBLE：控件不可见，但布局中仍然保留了控件所占用的显示空间。

(3) GONE：控件不可见且隐藏，控件不占用布局显示空间。

第15行用setChecked()方法设置toggleButton1控件的状态，默认为false，代表关状态；true为开状态。

第17～18行设置toggleButton1处于开、关状态时对应的显示文字。英文版默认为ON、OFF，中文版早期的版本对应"打开"和"关闭"，新版本对应"开启"和"关闭"。为避免版本差异导致的文字差异，建议用户自行设置开、关状态对应显示文字。

第20行用于变更toggleButton1的背景，不同的开、关状态可对应不同的背景图片。

第21行将toggleButton1的文字改为红色。

第22～40行定义了OnCheckedChangeListener监听器，其中，第29～35行判断toggleButton1的状态来变更imageView1绑定的图片文件，实现开、关灯效果。这几行代码也可以简化为第37行的三元表达式。问号前面是布尔表达式，条件为真就选择冒号前的图片文件，否则选择冒号后边的图片文件。

第38行和第47行的Log.i()方法用于演示toggleButton1的OnCheckedChangeListener和OnClickListener的触发先后顺序，同时还可演示调用callOnClick()方法和performClick()方法时的差异。调用callOnClick()方法只会触发OnClickListener监听器。调用performClick()方

法会依次触发 OnCheckedChangeListener 和 OnClickListener 监听器。performClick()方法等效于手动单击了 toggleButton1 按钮,callOnClick()方法可视为 performClick()方法的简化版。

第 66 行获取 toggleButton1 的相关属性值。

4.19 Switch

Switch 控件是带状态的开关控件,功能上基本与 ToggleButton、CheckedTextView 类似,只是外观上略有差异,如图 4-76 所示。控件的文字显示在左侧,"开""关"按钮图形在右侧;常用于 App 设置页面的开关型属性设置。

图 4-76　Switch 运行结果

```
【main.xml】
01  <?xml version = "1.0" encoding = "UTF - 8"?>
02  < LinearLayout xmlns:android = "http://schemas.android.com/apk/res/android"
03
04      android:layout_width = "match_parent"
05      android:layout_height = "match_parent"
06      android:orientation = "vertical">
07
08      < TextView
09          android:id = "@ + id/textView1"
10          android:layout_width = "match_parent"
11          android:layout_height = "wrap_content"
12          android:text = "TextView"
13          android:textColor = "@android:color/holo_red_dark"
14          android:textSize = "24sp" />
15
16      < Switch
17          android:id = "@ + id/switch1"
18          android:layout_width = "match_parent"
19          android:layout_height = "wrap_content"
20          android:text = "@string/switch1" />
21
22      < Switch
23          android:id = "@ + id/switch2"
24          android:layout_width = "match_parent"
25          android:layout_height = "wrap_content"
26          android:text = "@string/switch2"
27          android:theme = "@android:style/Theme.Holo.Light" />
28  </LinearLayout >
```

布局文件中放置了两个 Switch,第 27 行添加 android:theme 属性,其值对应老版本 Switch 外观。Switch 运行结果如图 4-76 所示。

【FirstActivity.java】
```
01    public class FirstActivity extends Activity
02    {
03        @Override
04        public void onCreate(Bundle savedInstanceState)
05        {
06            super.onCreate(savedInstanceState);
07            setContentView(R.layout.main);
08            TextView textView = (TextView) findViewById(R.id.textView1);
09            Switch switch1 = (Switch) findViewById(R.id.switch1);
10            switch1.setTextOn("打开");
11            switch1.setTextOff("关闭");
12            switch1.setShowText(true);        //打开开关上的文字显示,默认关闭。API 21
                                                //以上支持,演示局限性
13
14            switch1.setOnCheckedChangeListener(new CompoundButton.OnCheckedChangeListener()
15            {
16                @Override
17                public void onCheckedChanged(CompoundButton buttonView, boolean isChecked)
18                {
19                    textView.setText(isChecked?"打开开关":"关闭开关");
20                }
21            });
22        }
23    }
```

特殊一点的命令是第 12 行的 setShowText() 方法,如果设为 true,运行时 Switch 图形上会显示 setTextOn() 或 setTextOff() 方法指定的文字。

第 19 行调用监听器中布尔型变量 isChecked 值来决定 textView 中显示的文字。

setShowText 设为 true 的 Switch 运行结果如图 4-77 所示。

图 4-77　setShowText 设为 true 的 Switch 运行结果

视频讲解

4.20　AutoCompleteTextView

AutoCompleteTextView 控件是为了弥补 Spinner 控件中用户不能添加或修改下拉列表选项的局限。当用户输入字符时就以下拉列表的方式显示匹配的字符串。AutoCompleteTextView 在布局中的属性与 EditText 控件类似,重点看 Java 代码。

【FirstActivity.java】
```
01    public class FirstActivity extends Activity
02    {
03        @Override
04        public void onCreate(Bundle savedInstanceState)
05        {
06            super.onCreate(savedInstanceState);
```

```java
07      setContentView(R.layout.main);
08
09      AutoCompleteTextView autoCompleteTextView1 = (AutoCompleteTextView) findViewById
        (R.id.autoCompleteTextView1);
10      TextView textView1 = (TextView) findViewById(R.id.textView1);
11
12      String[] str = {"BeiJing", "ChongQing", "ChangSha", "Nan chang", "湖北", "湖
        南", "Si Chuan", "云南"};
13      ArrayAdapter<String> adapter = new ArrayAdapter<String>(this, android.R.
        layout.simple_dropdown_item_1line, str);//初始化适配器,指定下拉列表类型,
                                                //并绑定字符串数组
14
15      //ArrayAdapter<String> adapter = new ArrayAdapter<String>(this, android.
        R.layout.simple_spinner_item, str);      //simple_spinner_item 类型下拉列表
16
17      //ArrayAdapter<String> adapter = new ArrayAdapter<String>(this, android.
        R.layout.simple_spinner_dropdown_item, str);   //单选按钮类型下拉列表
18
19      autoCompleteTextView1.setAdapter(adapter);     //绑定适配器
20      autoCompleteTextView1.setThreshold(1);         //设置匹配字符数阈值,默认
                                                       //输入两个字符后开始匹配
21      autoCompleteTextView1.setCompletionHint("请选择匹配的选项");
                                                       //设置提示信息
22
23      //单击匹配下拉选项时触发
24      autoCompleteTextView1.setOnItemClickListener(new AdapterView.OnItemClickListener()
25      {
26          @Override
27          public void onItemClick(AdapterView<?> parent, View view, int position,
            long id)
28          {
29              textView1.setText("选择项: " + parent.getItemAtPosition(position).
                toString() + "\n\nparent: " + parent.toString() + "\n\nview: " +
                view.toString() + "\n\nposition: " + position + "\n\nid:" + id +
                "\n\nview的值: " + ((TextView) view).getText());
30              //讲解((TextView) view).getText()
31              Log.i("xj", "OnItemClickListener");
32          }
33      });
34
35      //单击 autoCompleteTextView1 控件时触发
36      autoCompleteTextView1.setOnClickListener(new View.OnClickListener()
37      {
38          @Override
39          public void onClick(View v)
40          {
41              Log.i("xj", "OnClickListener");
42          }
43      });
44
45      //不触发
```

```java
46            autoCompleteTextView1.setOnItemSelectedListener(new AdapterView.OnItemSelec-
              tedListener()
47            {
48                @Override
49                public void onItemSelected(AdapterView<?> parent, View view, int position,
                  long id)
50                {
51                    Log.i("xj", "OnItemSelectedListener");
52                }
53                @Override
54                public void onNothingSelected(AdapterView<?> parent)
55                {

56
57                }
58            });
59        }
60    }
```

AutoCompleteTextView 可采用类似 Spinner 案例中的 3 种方式进行初始化赋值,同样可对下拉列表的风格进行设定或修改。下面重点讲解与 Spinner 不同的地方。

autoCompleteTextView1 默认要输入两个字符才弹出匹配字符串的下拉列表框,第 20 行的 setThreshold() 方法的作用是设置匹配字符数的阈值为 1,即只要输入字符就查找是否有匹配的选项,如果有匹配字符串就弹出下拉列表框。默认值为 2,当设定的值小于 1 时,自动设定为 1。

第 21 行的 setCompletionHint() 方法类似 Spinner 中的 setPrompt() 方法。注意查看弹出下拉列表框下方的提示。

Spinner 的 OnItemSelectedListener 监听器对 AutoCompleteTextView 无效,需要使用 OnItemClickListener 监听器来获取相关属性值。OnClickListener 监听器不是单击下拉列表框触发,而是单击 AutoCompleteTextView 文本框触发。大家在使用时一定要注意细节上的差异。

对于第 29 行中代码 ((TextView)view).getText(),可能有读者会奇怪为什么 view 转换的类型不是 AutoCompleteTextView,而是 TextView? 此时弹出的下拉列表框是其他一系列控件的组合,其中显示匹配选项的控件是 TextView,所以显示 view 的值要用 ((TextView)view).getText()。相应的 parent 是指所有匹配的控件组成的 DropDownListView,可用 position 进行定位。弹出的下拉列表框结构可使用 Android Studio 的 Layout Inspector 工具查看。运行 App,在 autoCompleteTextView1 中输入字符,当弹出匹配的下拉列表框时打开 Layout Inspector,在其中显示的界面中选择下拉列表框中对象查看名称及层级位置。使用 Layout Inspector 查看界面结构如图 4-78 界面。

AutoCompleteTextView 运行结果如图 4-79 所示。

【注】 AutoCompleteTextView 输入字符匹配原则:
(1) 匹配忽略大小写。
(2) 从第一个字符或空格后的第一个字符开始匹配。

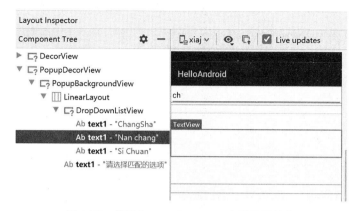

图 4-78 使用 Layout Inspector 查看界面结构

图 4-79 AutoCompleteTextView 运行结果

4.21 ScrollView 和 HorizontalScrollView

视频讲解

ScrollView 和 HorizontalScrollView 可对其内部控件实现上下或左右滑动效果。如果两个控件互相嵌套，对其共同控制的内部控件实现上下和左右滑动效果，但同一时刻只能在一个方向移动。本案例中添加了 12 个可滑动的按钮，这里不列出布局文件代码，可以从组件树来看各控件之间的位置关系。

由于 ScrollView 中只能包含一个控件，如果有多个控件需要放入 ScrollView 中，可先将其放在线性布局中，再将线性布局放到 ScrollView。ScrollView 中控件如图 4-80 所示，scrollView1 逻辑上只有线性布局一个控件，其余 12 个按钮都包含在线性布局中。

图 4-80 ScrollView 中控件

【FirstActivity.java】
```
01  public class FirstActivity extends Activity
02  {
03      @Override
04      public void onCreate(Bundle savedInstanceState)
05      {
06          super.onCreate(savedInstanceState);
07          setContentView(R.layout.main);
08
09          ScrollView scrollView1 = (ScrollView) findViewById(R.id.scrollView1);
10          Button button1 = (Button) findViewById(R.id.button1);
11          button1.setOnClickListener(new View.OnClickListener()
12          {
13              @Override
14              public void onClick(View v)
15              {
16                  if (button1.getText().toString().equals(getString(R.string.buttonLast)))
17                  {
18                      scrollView1.fullScroll(ScrollView.FOCUS_DOWN);
19                      button1.setText(R.string.buttonFirst);
20                      return;
21                  }
22                  scrollView1.fullScroll(ScrollView.FOCUS_UP);
23                  button1.setText(R.string.buttonLast);
24              }
25          });
26      }
27  }
```

运行程序后可用手指滑动以显示下方被遮挡的按钮。由于 button1 在 scrollView1 之外,因此是无法滑动的。

第 16 行判断 button1 的显示值是否为 strings.xml 文件中 buttonLast 对应的字符串,如果结果为 true,执行第 18 行将 scrollView1 向下滚动到最后一个控件,之后执行第 19 行修改按钮的显示文字,最后执行第 20 行 reture 命令结束按钮单击监听器的执行。

第 16~23 行等效于 if-else,如果第 16 行条件成立,则执行第 18~20 行,第 20 行是 return,也就不执行剩下的代码。如果条件不成立,则从第 22 行开始执行,相当于 else 代码块。现如今流行的趋势是优化 if-else,如果需要使用多个 if-else,常见的优化方式有:

(1) 使用 switch 替代,提高代码可读性。
(2) 使用 return 去除不必要的 else。
(3) 使用三元运算符。
(4) 其他方式根据具体情况可采用枚举、数组等方式优化。

ScrollView 运行结果如图 4-81 所示。

图 4-81　ScrollView 运行结果

ScrollView 还可以添加 android:scrollbarStyle 属性来设置右侧滑动条的风格,常用以下 4 个属性值。

(1) insideOverlay。
(2) insideOverlay。
(3) outsideOverlay。
(4) outsideInset。

4.22　TextClock

从 API 17 开始 TextClock 控件替代弃用的 DigitalClock 控件。可以使用 Date 或 Calendar 获取日期和时间并传递给 TextView,为什么还要使用 TextClock 控件呢？原因是 TextClock 既简单又能自动显示时间变化。

```
【main.xml】
01  <?xml version = "1.0" encoding = "UTF-8"?>
02  < LinearLayout xmlns:android = "http://schemas.android.com/apk/res/android"
03
04      android:layout_width = "match_parent"
05      android:layout_height = "match_parent"
06      android:orientation = "vertical">
07
08      < TextClock
09          android:layout_width = "wrap_content"
10          android:layout_height = "wrap_content"
11          android:textColor = "@android:color/holo_blue_dark"
12          android:textSize = "24sp" />
13
14      < TextClock
15          android:id = "@ + id/textClock1"
16          android:layout_width = "wrap_content"
17          android:layout_height = "wrap_content"
18          android:textColor = "#FF0000"
19          android:textSize = "24sp" />
20
21      < AnalogClock
22          android:layout_width = "wrap_content"
23          android:layout_height = "wrap_content" />
24  </LinearLayout >
```

布局文件中添加了两个 TextClock 控件和一个 AnalogClock 控件,后者已弃用,这里列出只是让读者了解 AnalogClock 的外观。

```
【FirstActivity.java】
01  public class FirstActivity extends Activity
02  {
03      private Chronometer chronometer = null;
```

```
04
05          @Override
06          public void onCreate(Bundle savedInstanceState)
07          {
08              super.onCreate(savedInstanceState);
09              setContentView(R.layout.main);
10
11              TextClock textClock1 = (TextClock)findViewById(R.id.textClock1);
12              textClock1.setTimeZone("GMT + 0800"); //设置时区
13              textClock1.setFormat12Hour("EEEE EE E \nyyyy MMMM MM M dd \nyyyy - MM - dd \nhh:
                mm:ss a");
14          }
15      }
```

第 12 行设置 textClock1 的时区,模拟器默认是 0 区,北京位于东 8 区,可使用 setTimeZone( )方法设定时区。参数字符串中的加减符号遵循东加西减的原则。后续的 4 位数中前两位数是时差,范围为[0,23],后两位数是分钟数,范围为[0,59]。北京时区除了表示为 0800 外,也可以表示为 08:00 或 08。

第 13 行设置 TextClock 的时间显示格式,默认只显示时间。可以使用指定的格式字符来显示相应的日期或时间字段。指定的字符位数不同也会导致显示内容不同,如 EEEE 显示星期一,EE 显示周一。其中的 ss 代表两位数显示的秒,显示的秒数会自动跳变,无须编程更新。

TextClock 运行结果如图 4-82 所示。

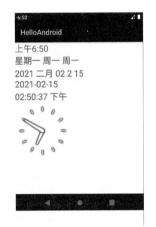

图 4-82　TextClock 运行结果

4.23　Chronometer

Chronometer 是作为计时器使用的控件,可当作闹钟、定时器使用。

```
【main.xml】
01  <?xml version = "1.0" encoding = "UTF - 8"?>
02  < LinearLayout xmlns:android = "http://schemas.android.com/apk/res/android"
03      android:layout_width = "match_parent"
04      android:layout_height = "match_parent"
05      android:orientation = "vertical">
06
07      < Chronometer
08          android:id = "@ + id/chronometer1"
09          android:layout_width = "wrap_content"
10          android:layout_height = "wrap_content"
11          android:layout_gravity = "center_horizontal" />
12
13      < LinearLayout
```

```xml
14          android:layout_width = "wrap_content"
15          android:layout_height = "wrap_content">
16
17      <Button
18          android:layout_width = "wrap_content"
19          android:layout_height = "wrap_content"
20          android:onClick = "onStart"
21          android:text = "开始计时显示" />
22
23      <Button
24          android:layout_width = "wrap_content"
25          android:layout_height = "wrap_content"
26          android:onClick = "onStop"
27          android:text = "停止计时显示" />
28
29      <Button
30          android:layout_width = "wrap_content"
31          android:layout_height = "wrap_content"
32          android:onClick = "onReset"
33          android:text = "重置计时器" />
34  </LinearLayout>
35
36  <TextView
37      android:id = "@ + id/textView1"
38      android:layout_width = "match_parent"
39      android:layout_height = "wrap_content" />
40 </LinearLayout>
```

布局文件中定义了一个 Chronometer、3 个按钮和一个 TextView。第 20 行是在布局文件中定义按钮的单击操作，会调用在 Java 中自定义的 onStart() 方法，运行结果与在 Java 中定义单击监听器一致。第 26 和 32 行同理。

【FirstActivity.java】
```java
01  public class FirstActivity extends Activity
02  {
03      Chronometer chronometer1;
04      TextView textView1;
05      Vibrator vibrator1;
06      boolean isStart = false;
07
08      @Override
09      public void onCreate(Bundle savedInstanceState)
10      {
11          super.onCreate(savedInstanceState);
12          setContentView(R.layout.main);
13
14          textView1 = (TextView) findViewById(R.id.textView1);
15          chronometer1 = (Chronometer) findViewById(R.id.chronometer1);
16          chronometer1.setBase(SystemClock.elapsedRealtime() - 5 * 1000);
```

```java
17          chronometer1.setFormat("计时：% s");
18
19          //计时监听事件,监听计时器时间的变化
20          chronometer1.setOnChronometerTickListener(new Chronometer.OnChronometerTi-
            ckListener()
21          {
22              @Override
23              public void onChronometerTick(Chronometer chronometer)
24              {
25                  String time = chronometer.getText().toString();
26                  if (time.equals("计时：00:10"))   //判断到指定时间时弹出 Toast
27                  {
28                      textView1.append("\n 闹钟响了!重要的事情响 3 遍!");
29
30                      vibrator1 = (Vibrator) getSystemService(VIBRATOR_SERVICE);
                        //获取振动服务
31                      if(vibrator1.hasVibrator()) //检查是否有振动器
32                      {
33                          //需要在 AndroidManifest 中注册振动权限,否则
                            //调用 vibrate()方法抛出异常
34                          try
35                          {
36                              vibrator1.vibrate(createWaveform(new long[]{100,
                                3000, 2000, 1000}, -1));
37                              isStart = true;
38                          } catch (Exception e)
39                          {
40                              textView1.append("出错了." + e.toString());
41                          }
42                      }
43                  }
44                  //textView1.append("\n 闹钟时间未到");
45              }
46          });
47
48          //判断运行此 App 的 Android 版本号
49          if (Build.VERSION.SDK_INT >= Build.VERSION_CODES.N)
50          {
51              //chronometer1.setCountDown(true); //倒计时,需要是 API 24,时间前会出现-号
52          }
53      }
54
55      //开始显示变动的时间
56      public void onStart(View view)
57      {
58          chronometer1.start();
59          textView1.append("\n 开始计时显示");
60      }
61
62      //停止显示计时
63      public void onStop(View view)
64      {
65          chronometer1.stop();
```

```
66          if (isStart)
67              vibrator1.cancel();    //停止振动
68          isStart = false;
69          textView1.append("\n 停止计时显示");
70      }
71
72      //重置计时器
73      public void onReset(View view)
74      {
75          chronometer1.setBase(SystemClock.elapsedRealtime());//设置计时器的起始时间
76          textView1.append("\n 时间重置到" + SystemClock.elapsedRealtime());
77      }
78  }
```

第 16 行设置计时器的基准时间,代码中将当前时间减去 5000ms,代表计时时间已经流逝 5000ms。

第 17 行设置计时器显示文字和格式,也是后续用于判断定时时间到期的字符串格式。

第 20 行定义了 OnChronometerTickListener 监听器,每过一秒触发一次。

第 30 行获取 Android 设备的振动服务。

第 31 行判断 Android 设备是否有振动器。

第 36 行调用振动的 vibrate() 方法并制定振动持续时间和间隔时间,振动需要获取权限才能使用,需要在 AndroidManifest.xml 中添加如下权限:

< uses - permission android:name = "android.permission.VIBRATE" />

如果没有设置权限,运行 vibrate() 方法将抛出异常,所以本行代码需要 try-catch 捕获异常。createWaveform() 方法的第一个参数 timings 是 long 型数组,new long[]{100,3000,2000,1000}数组的含义为等待 100ms,振动 3000ms,等待 2000ms,振动 1000ms。createWaveform() 方法最后一个参数 repeat 设为-1,代表之前的振动方式不重复执行。repeat 的取值范围为-1 到 long 数组长度减 1,当取值非-1 时,从 repeat 值作为索引的数组位置开始剩余部分的循环。开始循环后就会一直执行,直至清理项目或者调用 vibrator1.cancel() 为止。long 型数组长度一般大于 2 即可,可根据需要设定数组长度和数值,数组中数值遵循间歇、振动、间歇、振动的顺序往后排列。

第 37 行将自定义变量 isStart 设为 true,标记 vibrator1 已经获取权限并振动。设置此变量的目的是防止未获取振动服务就调用 vibrator1.cancel() 导致程序异常。

第 51 行 setCountDown 方法设为 true 时计时器变为倒计时,由于此命令需要 API 24 才支持,所以加入了 Android 版本判定。

从第 56~77 行定义布局文件中 3 个按钮的自定义方法,分别调用 chronometer1 的 start()、stop()和 setBase()方法。一旦执行 setBase()方法,计时器就清零并开始计时,无论是否执行 start()和 stop()方法(两个方法只是决定时间变动是否显示)。

Chronometer 运行结果如图 4-83 所示。

图 4-83　Chronometer 运行结果

4.24 AlertDialog

AlertDialog 对话框允许自定义弹出对话框的内容,实现比之前讲解的几种对话框更丰富的功能。

4.24.1 带默认按钮的 AlertDialog

【FirstActivity.java】
```java
01  public class FirstActivity extends Activity
02  {
03      EditText editText1;
04
05      @Override
06      public void onCreate(Bundle savedInstanceState)
07      {
08          super.onCreate(savedInstanceState);
09          setContentView(R.layout.main);
10          editText1 = (EditText) findViewById(R.id.editText1);
11          Button button1 = (Button) findViewById(R.id.button1);
12          button1.setOnClickListener(new View.OnClickListener()
13          {
14              @Override
15              public void onClick(View v)
16              {
17                  showAlertDialog();
18              }
19          });
20      }
21
22      void showAlertDialog()
23      {
24          Builder builder = new AlertDialog.Builder(this);      //使用对话框默认主题
                                                                  //初始化 builder
25          builder.setTitle("对话框 Title");                      //设置标题
26          builder.setMessage("对话框 Message");                  //设置提示信息
27          builder.setIcon(R.drawable.icon);                      //设置图标
28
29          //设置"确定"按钮
30          builder.setPositiveButton("确定", new DialogInterface.OnClickListener()
31          {
32              @Override
33              public void onClick(DialogInterface dialog, int which)
34              {
35                  editText1.setText("单击了对话框的"确定"按钮,which = " + which);
                    //数据交互
36              }
37          });
```

```
38
39                //设置"取消"按钮
40                builder.setNegativeButton("取消", new DialogInterface.OnClickListener()
41                {
42                    @Override
43                    public void onClick(DialogInterface dialog, int which)
44                    {
45                        editText1.setText("单击了对话框的"取消"按钮,which = " + which);
46                    }
47                });
48
49                //设置"中间"按钮
50                builder.setNeutralButton("中间", new DialogInterface.OnClickListener()
51                {
52                    @Override
53                    public void onClick(DialogInterface dialog, int which)
54                    {
55                        editText1.setText("单击了对话框的"中间"按钮,which = " + which);
56                    }
57                });
58                builder.setCancelable(false);              //对话框不允许单击背景消失
59                AlertDialog alertDialog1 = builder.create();  //使用builder创建对话框
60                alertDialog1.show();                        //显示对话框
61                //alertDialog.cancel();//对话框关闭,注意与builder.setCancelable(false)的区别
62            }
63  }
```

程序单击按钮后执行第 17 行代码,调用自定义方法 showAlertDialog()。

第 22～62 行是自定义 showAlertDialog()方法。调用 AlertDialog()的顺序是先定义 Builder;对 Builder 调用相关方法设定属性;定义默认按钮及单击监听器;调用 Builder 的 create()方法生成 AlertDialog;最后调用 AlertDialog 的 show()方法显示对话框。

第 24 行初始化 builder,此时 builder 拥有 AlertDialog 对话框的默认主题。

第 25 行调用 setTitle()方法设置对话框的标题。

第 26 行调用 setMessage()方法设置对话框的提示信息。

第 27 行调用 setIcon()方法设置对话框图标。

第 30～37 行添加 PositiveButton 按钮,其中 setPositiveButton()方法的第一个参数是按钮的显示文本。为了便于阅读代码,此按钮文本尽量采用确定语气的文字,如"是""确定"等。第二个参数是按钮的单击监听器。单击此按钮将自动关闭对话框,同时执行第 35 行代码,在文本输入框 editText1 中显示信息。变量 which 是 PositiveButton 按钮的 int 型返回值。

第 40～57 行以同样的方式设置"取消"和"中间"按钮。不同的 Android 版本 3 个按钮的外观和位置会有差异。每个对话框可以设置 1～3 个对话框默认按钮,也可以不设置按钮。

第 58 行的作用是单击对话框以外的区域,对话框仍然显示在最上层。默认值为 true,单击对话框以外的区域,对话框将被遮挡。

第 59 行调用 builder 的 create()方法生成 alertDialog1 对话框。

第 60 行调用 alertDialog1 的 show()方法显示对话框。

带默认按钮的 AlertDialog 运行结果如图 4-84 和图 4-85 所示。

图 4-84　弹出带默认按钮的 AlertDialog　　图 4-85　返回对话框按钮信息

4.24.2　列表的 AlertDialog

AlertDialog 除了默认主题的对话框形式外，还支持多种类型的对话框。本案例讲解列表风格的 AlertDialog。

```
【FirstActivity.java】
01  public class FirstActivity extends Activity
02  {
03      EditText editText1;
04      String[] str = {"Java", "C++", "Python", "C#"};
05
06      @Override
07      public void onCreate(Bundle savedInstanceState)
08      {
09          super.onCreate(savedInstanceState);
10          setContentView(R.layout.main);
11          editText1 = (EditText) findViewById(R.id.editText1);
12          Button button = (Button) findViewById(R.id.button1);
13          button.setOnClickListener(new View.OnClickListener()
14          {
15              @Override
16              public void onClick(View v)
17              {
18                  showAlertDialog(); //显示对话框
19              }
20          });
21      }
```

```
22
23      void showAlertDialog()
24      {
25          Builder builder = new AlertDialog.Builder(this);
26          builder.setIcon(R.drawable.icon);
27          builder.setTitle("请选择开发语言");
28          //设置为列表
29          builder.setItems(str, new DialogInterface.OnClickListener()
30          {
31              @Override
32              public void onClick(DialogInterface dialog, int which)
33              {
34                  editText1.setText("你选择的开发语言是: " + str[which]);
35              }
36          });
37          AlertDialog alertDialog1 = builder.create();
38          alertDialog1.show();
39      }
40  }
```

本案例的大多数代码与4.24.1节案例的代码类似,不同点在第29行,setItems()方法将对话框设置为选项列表风格。第一个参数是用于填充下拉选项列表的字符串数组str,str数组在第4行定义。第二个参数为单击监听器,处理单击选项后的触发事件。

第34行的OnClick()方法以变量which(单击选项列表返回的选中项索引值)为字符串数组str的索引,获取对应选中列表中的文本,并将其传递给editText1。

列表风格的AlertDialog运行结果如图4-86和图4-87所示。

图4-86　弹出列表风格的AlertDialog　　　图4-87　返回列表风格的AlertDialog的选择信息

4.24.3　单选的 AlertDialog

单选的 AlertDialog 是指弹出的对话框显示单选按钮风格的 AlertDialog。

【FirstActivity.java】
```
01  public class FirstActivity extends Activity
02  {
03      EditText editText1;
04      String[] str = {"Java", "C++", "Python", "C#"};
05      int num;
06
07      @Override
08      public void onCreate(Bundle savedInstanceState)
09      {
10          super.onCreate(savedInstanceState);
11          setContentView(R.layout.main);
12
13          editText1 = (EditText) findViewById(R.id.editText1);    //获取对象
14          Button button = (Button) findViewById(R.id.button1);    //获取对象
15          button.setOnClickListener(new View.OnClickListener()
16          {
17              @Override
18              public void onClick(View v)
19              {
20                  showAlertDialog();                              //显示对话框
21              }
22          });
23      }
24
25      void showAlertDialog()
26      {
27          Builder builder = new AlertDialog.Builder(this);
28          builder.setIcon(R.drawable.icon);
29          builder.setTitle("请选择开发语言");
30
31          builder.setSingleChoiceItems(str, -1, new DialogInterface.OnClickListener()
32          {
33              @Override
34              public void onClick(DialogInterface dialog, int which)
35              {
36                  num = which;
37                  editText1.setText("你选择的开发语言是:" + str[which]);
38                  //dialog.cancel();
39              }
40          });
41
42          builder.setPositiveButton("确定", new DialogInterface.OnClickListener()
43          {
44              @Override
45              public void onClick(DialogInterface dialog, int which)
46              {
47                  editText1.setText("你确定的开发语言是:" + str[num]);
48                  Log.i("xj", "which = " + which + ", num = " + num);
49              }
```

```
50                });
51            AlertDialog alertDialog = builder.create();
52            alertDialog.show();
53        }
54    }
```

第 4 行定义用于单选列表的字符串数组 str。

第 5 行的变量 num 用于记录单选风格对话框中选中项的索引值。

关键代码在第 31 行,setSingleChoiceItems()方法将对话框设置为单选按钮风格。其中第二个参数 checkedItem 是设置单选按钮列表选中项的索引值,如果是 -1 则所有单选按钮都设置为未选。

第 34 行的变量 which 是单选按钮列表的索引值,也是选中项对应字符串数组 str 的索引值。

第 45 行的变量 which 是 AlertDialog 对话框中按钮被单击时返回按钮所对应的 int 型数值,与第 34 行的 which 含义是不同的。

如果去掉第 38 行的注释,在弹出的对话框中选中单选项后,cancel()方法将关闭对话框而无须再单击"确定"按钮,也就不会执行第 42~50 行的代码。

单选的 AlertDialog 运行结果如图 4-88 所示。

图 4-88　单选的 AlertDialog 运行结果

4.24.4　复选的 AlertDialog

复选的 AlertDialog 是指弹出的对话框显示复选框风格的 AlertDialog。

```
【FirstActivity.java】
01    public class FirstActivity extends Activity
02    {
03        EditText editText1;
04        String[] str = {"Java", "C++", "Python", "C#"};
05        boolean[] flags = {true, false, true, false};        //布尔数组,true 代表选中
06
07        @Override
08        public void onCreate(Bundle savedInstanceState)
09        {
10            super.onCreate(savedInstanceState);
11            setContentView(R.layout.main);
12
13            editText1 = (EditText) findViewById(R.id.editText1);
14            Button button1 = (Button) findViewById(R.id.button1);
15            button1.setOnClickListener(new View.OnClickListener()
16            {
17                @Override
18                public void onClick(View v)
```

```
19          {
20              showAlertDialog();
21          }
22      });
23  }
24
25  void showAlertDialog()
26  {
27      Builder builder = new AlertDialog.Builder(this);
28      builder.setIcon(R.drawable.icon);
29      builder.setTitle("请选择开发语言");
30      builder.setMultiChoiceItems(str, flags, new DialogInterface.OnMultiChoiceCl-
            ickListener()
31      {
32          @Override
33          public void onClick(DialogInterface dialog, int which, boolean isChecked)
34          {
35              flags[which] = isChecked;         //哪一项被选中
36          }
37      });                                       //设置复选列表
38
39      builder.setPositiveButton("确定", new DialogInterface.OnClickListener()
40      {
41          @Override
42          public void onClick(DialogInterface dialog, int which)
43          {
44              String temp = "你选择的开发语言是：";
45              for (int i = 0; i < flags.length; i++)
46              {
47                  if (flags[i])
48                      temp += str[i] + ";";
49              }
50              editText1.setText(temp);          //返回选中的复选框内容
51          }
52      });                                       //设置"确定"按钮
53      AlertDialog alertDialog1 = builder.create();
54      alertDialog1.show();
55  }
56 }
```

第 4 行定义用于复选框列表的字符串数组 str。

第 5 行的布尔型数组 flags 定义了 str 数组在复选框列表中的选中状态，true 为选中。flags 数组的长度应与 str 数组长度相同。如果 flags 数组长度小于 str 数组长度，程序运行到第 30 行绑定两个数组到 builder 时会出错。如果 flags 数组长度大于 str 数组长度，程序运行到第 45 行取 flags 数组的长度，执行到第 48 行可能会因数组索引超出 str 数组范围而出错，只要将第 45 行的 flags.length 改成 str.length 程序就能正常运行。

第 30 行的 setMultiChoiceItems() 方法将 AlertDialog 设置为复选框列表。

第 35 行将复选项的变动值（true 或 false）同步到 flags 数组的对应项。

第 45 行使用 for 循环遍历数组 flags,状态为 true 的索引项返回 str 数组的对应值。

第 50 行将对话框的返回结果传递给 editText1。

第 53~54 行调用 builder 的 create 方法生成 alertDialog1,再调用 show 方法显示对话框。

4.24.5 自定义控件

目前为止已经介绍了多种对话框,这些对话框的内置控件都是定制好的。本案例讲解自定义对话框中的控件。为便于讲解,对话框中的控件只加入 EditText 和 Button,读者理解设计方式后可自行添加所需控件。

【FirstActivity.java】
```
01   public class FirstActivity extends Activity
02   {
03       EditText editText1;
04       @Override
05       public void onCreate(Bundle savedInstanceState)
06       {
07           super.onCreate(savedInstanceState);
08           setContentView(R.layout.main);
09
10           editText1 = (EditText) findViewById(R.id.editText1);
11           Button button1 = (Button) findViewById(R.id.button1);
12           button1.setOnClickListener(new View.OnClickListener()
13           {
14               @Override
15               public void onClick(View v)
16               {
17                   showAlertDialog(10);       //显示动态初始化对话框
18                   //showAlertDialog(20);     //显示 login.xml 对话框(不带值返回)
19                   //showAlertDialog(30);     //显示 login.xml 对话框(带值返回)
20               }
21           });
22       }
23
24       void showAlertDialog(int id)
25       {
26           Builder builder = new AlertDialog.Builder(this);
27           switch (id)
28           {
29               case 10:                       //直接初始化一个输入框到弹出对话框
30                   EditText editTextName = new EditText(this);
                     //动态初始化一个输入框
31                   builder.setView(editTextName);  //输入框绑定到对话框
32                   builder.setTitle("请输入您的姓名:");
33                   builder.setPositiveButton("确定", new DialogInterface.OnClickListener()
34                   {
35                       @Override
```

```
36                    public void onClick(DialogInterface dialog, int which)
37                    {
38                            editText1.setText("您的姓名是: " + editTextName.getText());
39                    }
40                });
41                break;
42       case 20:    //方法一:将login.xml布局文件绑定到弹出对话框
43                LayoutInflater inflater20 = getLayoutInflater();
                  //LayoutInflater是用来找layout目录下的布局文件,并将之实例化
44                View viewLogin20 = inflater20.inflate(R.layout.login, (ViewGroup)
                  findViewById(R.id.linearLayoutLogin));
45                builder.setIcon(R.drawable.icon).setTitle("请输入您的姓名: ").
                  setView(viewLogin20).setPositiveButton("确定", null).
                  setNegativeButton("取消", null);
46                break;
47       case 30:    //方法二:将login.xml布局文件绑定到弹出对话框
48                LayoutInflater inflater30 = getLayoutInflater();
49                View viewLogin30 = inflater30.inflate(R.layout.login, (ViewGroup)
                  findViewById(R.id.linearLayoutLogin));
50                builder.setTitle("请输入您的姓名: ");
51                builder.setView(viewLogin30);
52                builder.setPositiveButton("确定", new DialogInterface.OnClickListener()
53                {
54                    @Override
55                    public void onClick(DialogInterface dialog, int which)
56                    {
57                            editText1.setText("您的姓名是: " + ((EditText) viewLogin30.
                            findViewById(R.id.editTextName)).getText());
58                            //必须在findViewById()方法之前写上viewLogin30,
                            //否则默认为main.xml的view
59                    }
60                });      //设置"确定"按钮
61
62                break;
63       default:
64                break;
65       }
66       AlertDialog alertDialog = builder.create(); //创建对话框
67       alertDialog.show();
68   }
69 }
```

第17~19行通过自定义showAlertDialog()方法的不同参数调用不同形式的自定义对话框。

当参数为10时执行第29行开始的代码,实现动态添加控件,常用于对布局要求不高的场景。第30行使用EditText的构造方法生成一个文本输入框editTextName。第31行将生成的editTextName文本输入框添加到builder中。第32行添加对话框的标题。第33行添加"确定"按钮,按钮的单击监听器将对话框中的editTextName内容传回main.xml布局

文件中的 editText1 中。

当参数为 20 时执行第 42 行开始的代码。此时调用已经设计好的布局文件 login.xml。第 43 行声明 LayoutInflater 类型的变量 inflater20。第 44 行使用 inflate() 方法将 login.xml 中名为 linearLayoutLogin 的线性布局作为 View 填充到 viewLogin20。第 45 行是将 builder 的一系列方法放在一行中连续调用。相关按钮的监听器都设置为 null，单击这些按钮就只能执行默认的关闭对话框操作。

当参数为 30 时执行第 47 行开始的代码，是对方法一的完善，可以返回文本输入框中的文字内容。实现的方式是重写按钮单击监听器代码。在第 57 行的 findViewById() 方法前加了 viewLogin30 限定。第 49 行 viewLogin30 对应的是 login.xml 布局文件的 linearLayoutLogin。如果没有 viewLogin30 前缀，查找 R.id.editTextName 从 main.xml 中查询，找不到程序就会出错退出。如果两个文件中控件重名，没有 viewLogin30 前缀的 findViewById() 方法取回的就是 main.xml 布局文件中的控件。

带自定义控件 AlertDialog 运行结果如图 4-89 和图 4-90 所示。

图 4-89　弹出带自定义控件 AlertDialog

图 4-90　返回带自定义控件 AlertDialog 信息

【注】　Spinner、AutoCompleteTextView 和 AlertDialog 都支持类似的弹出框功能，使用时注意各自的优缺点和使用场景。

第 5 章　其他常用编程技术

本节汇集了 Android 应用开发中经常涉及的一些相关编程技术。

5.1　Intent

当从一个 Activity 调用另外的 Activity 时，需要使用 Intent。当使用 Service、Broadcast 或第三方 App 的功能时也需要使用 Intent。Intent 还负责调用或返回 Activity 时的数据传递。

5.1.1　Intent 的显式调用和隐式调用

Intent 的调用分为显示调用和隐式调用。

（1）显式调用：明确 Intent 要调用的 Activity 名称（AndroidManifest.xml 中 activity 标签的 android:name 或者 Java 中定义的 Activity 类名称）。

（2）隐式调用：不明确指定启动的 Activity，通过设置 Action、Data、Category，由系统根据 intent-filter 来筛选合适的 Activity。隐式调用的功能非常强大，可以调用系统的应用，如桌面、相机、浏览器等。

本案例设计了两个布局文件，分别取名 main.xml 和 second.xml。关键文件的代码如下：

```
【AndroidManifest.xml】
01  <?xml version = "1.0" encoding = "UTF-8"?>
02  <manifest xmlns:android = "http://schemas.android.com/apk/res/android"
03      package = "com.xiaj">
04
05      <application
06          android:icon = "@drawable/icon"
07          android:label = "@string/app_name"
08          android:theme = "@android:style/Theme.Holo.Light">
09          <activity
10              android:name = "com.xiaj.FirstActivity"
11              android:label = "@string/app_name">
12              <intent-filter>
13                  <action android:name = "android.intent.action.MAIN" />
14                  <category android:name = "android.intent.category.LAUNCHER" />
15              </intent-filter>
16          </activity>
17          <activity
18              android:name = "com.xiaj.SecondActivity"
19              android:label = "@string/app_name">
```

```
20              <intent-filter>
21                  <action android:name = "com.xiaj.SecondActivityAction" />
22                  <category android:name = "android.intent.category.DEFAULT" />
23              </intent-filter>
24          </activity>
25
26      </application>
27 </manifest>
```

新添加 Activity 时 Android Studio 会自动在 AndroidManifest.xml 文件中添加 activity 标签进行注册。如果调用未注册的 Activity 将导致运行出错。第一个添加的 Activity 会自动添加第 12~15 行的 intent-filter。其中,第 13 行定义当前 Activity 对应的 action,其值 android.intent.action.MAIN 代表当前 Activity 是 App 的首选运行 Activity。第 14 行的 android.intent.category.LAUNCHER 决定应用程序是否显示在程序列表中。第 13 行和第 14 行合在一起决定 App 启动后会优先启动的 Activity。如果多个 Activity 中都有第 12~15 行的代码,则按 Activity 在 AndroidManifest.xml 中出现的顺序决定首先启动的 Activity。第 20~23 行的 intent-filter 标签是为了讲解 Intent 的隐式调用。

【FirstActivity.java】
```
01  public class FirstActivity extends Activity
02  {
03      @Override
04      public void onCreate(Bundle savedInstanceState)
05      {
06          super.onCreate(savedInstanceState);
07          setContentView(R.layout.main);
08          Button button1 = (Button) findViewById(R.id.buttonFirst); //获取 id
09          button1.setOnClickListener(new View.OnClickListener()
10          {
11              @Override
12              public void onClick(View v)
13              {
14                  Intent intent = new Intent(); //Intent 对象
15                  //方法一
16                  //intent.setClass(FirstActivity.this, SecondActivity.class);
17                  //intent.setClass(getApplicationContext(), SecondActivity.class);
18
19                  //方法二
20                  //intent.setClassName(getApplicationContext(), "com.xiaj.SecondActivity");
21                  //intent.setClassName("com.xiaj", "com.xiaj.SecondActivity");
22
23                  //方法三
24                  //ComponentName componentName = new ComponentName(FirstActivity.this,
                    //"com.xiaj.SecondActivity");
25                  //intent.setComponent(componentName);
26
```

```
27                //方法四
28                //intent.setAction("com.xiaj.SecondActivityAction");
29
30                //方法五
31                //intent = new Intent("com.xiaj.SecondActivityAction");
32
33                //方法六
34                //intent.setData(Uri.parse("https://www.qq.com"));
35
36                startActivity(intent); //启动
37            }
38        });
39    }
40 }
```

代码中前 3 种方法属于显式调用,后 3 种方法属于隐式调用。其中,方法四和方法五必须要在 AndroidManifest.xml 中有 intent-filter 标签,其中的 action 和 category 标签也是不可缺少的,否则调用时也会出错。方法六调用后直接使用默认浏览器访问指定网站,这为程序的功能扩展提供了极大的便利。

【注】 当同时使用显式调用和隐式调用时,优先选择显式调用。

```
【SecondActivity.java】
01 public class SecondActivity extends Activity
02 {
03     @Override
04     protected void onCreate(Bundle savedInstanceState)
05     {
06         super.onCreate(savedInstanceState);
07         setContentView(R.layout.second);
08
09         TextView textView1 = (TextView) findViewById(R.id.textView1);
10         Intent intent = getIntent();         //获取当前 Intent 对象
11         ComponentName componentName = intent.getComponent();       //获取组件名称
12
13         textView1.setText("PackageName: " + componentName.getPackageName());
14         textView1.append("\nClassNam: " + componentName.getClassName());
15         textView1.append("\nShortClassName: " + componentName.getShortClassName());
16         textView1.append("\nAction: " + intent.getAction());
17     }
18 }
```

当程序转到 SecondActivity 时,可用上述代码获取 Intent 中相关信息。其中第 16 行的 getAction()方法需要 FirstActivity 中的方法四和方法五才会显示非 null 值。SecondActivity 运行结果如图 5-1 所示。

图 5-1　SecondActivity 运行结果

5.1.2 Intent 传值和取值

本案例演示从 FirstActivity 调用并传送数值到 SecondActivity,在 SecondActivity 运行后将新的数值传回 FirstActivity。本案例布局文件与 5.1.1 节案例类似,重点看 Java 文件。

视频讲解

【FirstActivity.java】
```
01  public class FirstActivity extends Activity
02  {
03      TextView textViewShow;
04      @Override
05      public void onCreate(Bundle savedInstanceState)
06      {
07          super.onCreate(savedInstanceState);
08          setContentView(R.layout.main);
09
10          EditText editTextUserName = (EditText) findViewById(R.id.editTextUserName);
11          EditText editTextPassword = (EditText) findViewById(R.id.editTextPassword);
12          Button button1 = (Button) findViewById(R.id.button1);
13          textViewShow = (TextView) findViewById(R.id.textViewshow);
14
15          button1.setOnClickListener(new View.OnClickListener()
16          {
17              @Override
18              public void onClick(View v)
19              {
20                  Intent intent = new Intent(getApplicationContext(), SecondActivity.class);
21                  //方法一:传送参数
22                  intent.putExtra("name", editTextUserName.getText().toString());
23                  intent.putExtra("password", editTextPassword.getText().toString());
24
25                  //方法二:传送对象
26                  Bundle bundle1 = new Bundle();
27                  bundle1.putString("name", "张三丰");
28                  bundle1.putInt("age", 80);
29                  bundle1.putStringArray("徒弟们", new String[]{"宋远桥", "俞莲舟"});
30                  intent.putExtras(bundle1);
31                  //当无须返回数据时使用
32                  //startActivity(intent);
33
34                  //需要返回数据时使用
35                  startActivityForResult(intent, 10); //启动 SecondActivity,查看相
                                                        //关组件信息
36              }
37          });
38      }
39
```

```
40          @Override
41          protected void onActivityResult(int requestCode, int resultCode, Intent data)
42          {
43              super.onActivityResult(requestCode, resultCode, data);
44              if (resultCode == 20)            //20 代表 SecondActivity 发回的数据
45              {
46                  textViewShow.append("\nrequestCode:" + requestCode + "\nresultCode:" +
                    resultCode + "\nSecondActivity 传回的密码为:" + data.getStringExtra
                    ("password") + "\nintent:" + data.getExtras());
47              }
48          }
49      }
```

第 20 行使用 Intent 的构造方法显式调用 SecondActivity。

第 22~23 行调用 putExtra()方法按 key-value 键值对方式设定用户名和密码并绑定到变量 intent。比较容易犯错的地方是从 EditText 使用 getText()方法获取的 Editable 对象一定要使用 toString()方法转换为 String 类型，否则接收端按 String 型获取传送值时只能得到 null。

第 26 行演示使用 Bundle 方式传值。与 putExtra()方法传值相比，Bundle 可以传送非 String 类型的对象或数组。

第 27 行使用 putString()方法绑定 String 型键值对，此处 key 与第 22 行的 key 都是 name，字符串"张三丰"将覆盖第 22 行的 value 值"张三"。

第 28 行使用 putInt()方法绑定 int 型数值，也可以使用 putString()方法传送 String 型数值，接收方再将 String 型转换为 int 型。

第 29 行使用 putStringArray()方法将 String 型数组绑定到 bundle1。键值对中的 key 也可以使用中文，如本行中的"徒弟们"。

第 30 行将 bundle1 作为对象绑定到 intent。

第 32 行使用 startActivity()方法启动 intent(如果去掉注释的话)，此时系统会运行 SecondActivity，并将相关数据传送给 SecondActivity。此时可以将数据从 FirstActivity 传递到 SecondActivity，但无法将数据回传给 FirstActivity。

如果要实现数据从 SecondActivity 回传给 FirstActivity，需将第 32 行的 startActivity()方法替换成第 35 行的 startActivityForResult()方法。startActivityForResult()方法除了具有 startActivity()的功能以外，还具有等待被调用的 SecondActivity()返回 intent 的功能。startActivityForResult()方法多了一个参数 requestCode。如果 FirstActivity 中发起多个调用 SecondActivity 的 intent，requestCode 用于区分是谁发起的调用 SecondActivity。

当 FirstActivity 使用 startActivityForResult()方法转到 SecondActivity，且 SecondActivity 使用 setResult()方法返回结果时，会自动调用第 41 行的 onActivityResult()方法。方法中第一个参数 requestCode 对应 startActivityForResult()方法中的 requestCode，resultCode 对应 SecondActivity 中 setResult()方法的参数 resultCode。如此就知道是谁发起了请求，是谁给了回应。

【SecondActivity.java】
```
01  public class SecondActivity extends Activity
02  {
03      @Override
04      protected void onCreate(Bundle savedInstanceState)
05      {
06          super.onCreate(savedInstanceState);
07          setContentView(R.layout.second);
08
09          TextView textView1 = (TextView) findViewById(R.id.textViewshow);
10          EditText editTextPassword = (EditText) findViewById(R.id.editTextPassword);
11          Button button2 = (Button) findViewById(R.id.button2);
12          Intent intent = getIntent();
13
14          //方法一
15          String name = intent.getStringExtra("name");
16          //方法二
17          String password = intent.getExtras().getString("password");
18          //int age = intent.getExtras().getInt("age");
19          //int age = intent.getExtras().getInt("age", 20);
20          int age = intent.getIntExtra("age", 20);
21          //String[] 徒弟们 = intent.getExtras().getStringArray("徒弟们");
22          String[] 徒弟们 = intent.getStringArrayExtra("徒弟们");
23
24          textView1.append("用户名：" + name);
25          textView1.append("\n密码：" + password);
26          textView1.append("\n年龄：" + age);
27          for (String 徒弟 : 徒弟们)
28          {
29              textView1.append("\n徒弟：" + 徒弟);
30          }
31
32          //以下方法是方法二的变形
33          Bundle bundle1 = intent.getExtras();
34          editTextPassword.setText(bundle1.getString("password"));
35          textView1.append("\n对象：" + bundle1);
36
37          button2.setOnClickListener(new View.OnClickListener()
38          {
39              @Override
40              public void onClick(View v)
41              {
42                  Intent intent = new Intent();
43                  intent.putExtra("password", editTextPassword.getText().toString());
44                  setResult(20, intent);
45                  finish();//关闭当前Activity
46              }
47          });
48      }
49  }
```

SecondActivity 可接收 FirstActivity 传来的键值对或 Bundle 对象，然后将处理结果回传给 FirstActivity。

第 15 行使用 getStringExtra()方法通过参数中的字符串 name 获取对应值。此种方式获取键值的方式最简单。

第 17 行通过 getExtras()方法的 getString()方法获取键值对中的值，效果与 getStringExtra()完全相同。

针对不同的数据类型或数组，SDK 提供了 getInt()或 getStringArrayExtra()等不同的方法。第 18～20 行代码效果基本相同。第 18 行的代码如果没有查到键值对 age，则返回默认值 0，而第 19 和 20 行会返回指定的默认值 20。

第 33 行使用 intent.getExtras()方法将返回结果传递给 bundle1，在连续调用键值对时可提高效率。

第 21 行和第 22 行的效果完全相同，与前面键值对的区别是 key 为中文。Java 是支持中文作为变量名的。

【注】 Java 标识符要求：
(1) Java 是大小写敏感的语言。
(2) Java 的保留字不能作为标识符。
(3) 不能以数字开头。
(4) 不能含有空格、@、#、—等非法字符。
(5) 能见名知义。

第 43 行在回传的 intent 中绑定 password 键值对。

第 44 行的 setResult()方法会将第一个参数 resultCode 和第二个参数 intent 回传给 FirstActivity 的 onActivityResult()方法。至此完成了参数值的传送和回传功能。

Activity 间传值运行结果如图 5-2～图 5-4 所示。

图 5-2　FirstActivity 传值

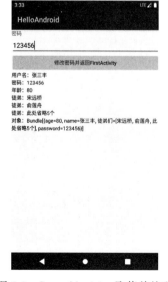

图 5-3　SecondActivity 取值并处理

本案例是不是很像之前讲过的 AlertDialog？外观上的区别就是 SecondActivity 不是对话框。如果在 AndroidManifest.xml 文件 SecondActivity 所在 Activity 标签中增加属性 **android:theme = "@style/Theme.AppCompat.Dialog"**，再次运行，SecondActivity 变成对话框外观。对话框风格的 Activity 运行结果如图 5-5 所示。

图 5-4　FirstActivity 取回处理值　　　图 5-5　对话框风格的 Activity 运行结果

5.2　Activity

布局文件实现 Android 应用程序的界面设计，而 Activity 作为 Android 的应用组件，通过 Java 代码设计实现 UI 交互功能。一个 App 通常由一个或多个 Activity 组成。

5.2.1　系统状态栏、标题栏和导航栏

视频讲解

默认每个 Activity 的界面都会显示系统状态栏、标题栏和导航栏。对于某些应用，需要将以上三者部分或全部隐藏。如果要实现 Activity 调用时就不显示标题栏，最简单的办法是在 AndroidManifest.xml 文件的 application 或 activity 标签中加入以下代码：

`android:theme = "@android:style/Theme.NoTitleBar"`

上述代码加在 application 标签中，表示所有的 Activity 都不显示标题栏。上述代码加在 activity 标签中，表示当前 activity 不显示标题栏。如果在上述节点位置加入：

`android:theme = "@android:style/Theme.NoTitleBar.Fullscreen"`

代表相应的 Activity 不显示系统状态栏和标题栏。其他几种隐藏系统状态栏和标题栏的方法都是在 Activity 的 Java 文件中进行设置的。

【FirstActivity.java】
```
01  public class FirstActivity extends Activity
02  {
03      @Override
04      public void onCreate(Bundle savedInstanceState)
05      {
06          super.onCreate(savedInstanceState);
07          //requestWindowFeature(Window.FEATURE_NO_TITLE);
08          setContentView(R.layout.main);
09
10          //getWindow().setFlags(WindowManager.LayoutParams.FLAG_FULLSCREEN,WindowManager.
              //LayoutParams.FLAG_FULLSCREEN);
11
12          Button button1 = (Button) findViewById(R.id.button1);
13          Button button2 = (Button) findViewById(R.id.button2);
14
15          //方案五：隐藏/显示标题栏
16          button1.setOnClickListener(new View.OnClickListener()
17          {
18              @Override
19              public void onClick(View v)
20              {
21                  ActionBar actionBar = getActionBar();
22                  if (button1.getText().toString().equals(getString(R.string.titleVisible)))
23                  {
24                      actionBar.show();
25                      button1.setText(getString(R.string.titleInvisible));
26                      return;
27                  }
28                  actionBar.hide();
29                  button1.setText(getString(R.string.titleVisible));
30              }
31          });
32
33          //隐藏系统状态栏和导航栏
34          View decorView = getWindow().getDecorView();
35          button2.setOnClickListener(new View.OnClickListener()
36          {
37              @Override
38              public void onClick(View v)
39              {
40
41                  if (button2.getText().toString().equals(getString(R.string.statusVisible)))
42                  {
43                      //decorView.setSystemUiVisibility(View.VISIBLE);
44                      decorView.setSystemUiVisibility(View.SYSTEM_UI_FLAG_VISIBLE);
45                      button2.setText(getString(R.string.statusInvisible));
46                  } else
```

```
47                           {
48                               decorView.setSystemUiVisibility(View.SYSTEM_UI_FLAG_FULLSCREEN |
                                 View.SYSTEM_UI_FLAG_HIDE_NAVIGATION | View.SYSTEM_UI_FLAG_
                                 IMMERSIVE);
49                               button2.setText(getString(R.string.statusVisible));
50                           }
51                       }
52                   });
53           }
54       }
```

Activity 内各 View 间关系如图 5-6 所示。因此显示了 Activity 内各 View 的大致关系,具体逻辑关系可在 Android Studio 运行程序后选择 Android Studio 右下角的 Layout Inspector 选项卡,在弹出的界面中查看各个 View 之间的关系嵌套,如图 5-7 所示。每个 Activity 对应一个 Window。DecorView 是 Window 中的顶层视图,StatusBar 是系统状态栏,ActionBar 是标题栏,Navigation 是导航栏,ContentView 就是第 8 行 setContentView()方法调入的布局文件形成的 View。隐藏或显示系统状态栏、标题栏和导航栏就是对 Activity 中的相应 View 执行隐藏或显示操作。

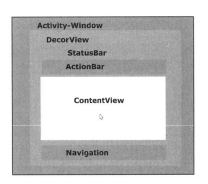

图 5-6　Activity 内各 View 间关系

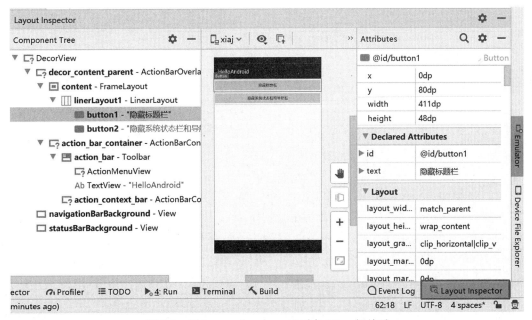

图 5-7　用 Layout Inspector 查看各 View 间关系

如果去掉第 7 行的注释,当前 Activity 将不显示标题栏。此命令必须在第 8 行 setContentView()方法之前,否则程序运行出错。加上之前介绍的隐藏标题栏的方法,如果再执行第 24 行或第 28 行的 actionBar 操作时会因标题栏设置冲突而出错。上述介绍的几种隐藏系统状态栏和标题栏的方法都无法在 Activity 启动以后再进行动态变更。

第 10 行的代码可放置在 setContentView() 方法前后任意位置(有些特殊属性需将此方法放在 setContentView() 方法前),通过 getWindow() 方法获取 Window,然后用 setFlags() 方法设置标记来实现全屏效果。

第 16~31 行定义 button1 的单击监听器,通过第 22 行判断按钮文字内容来执行标题栏的显示或隐藏操作。第 21 行获取标题栏对象并赋予 actionBar。第 24 行 actionBar 的 show() 方法用来显示标题栏,第 28 行的 hide() 方法用来隐藏标题栏。

第 34 行的 getWindow() 方法获取当前 Activity 对应的 Window,getDecorView() 方法获取当前 Window 包含的 View。第 43 行和第 44 行效果是一样的,利用 decorView 实例的 setSystemUiVisibility() 方法设置 UI 的相关对象是否显示。其中,常量的含义为:

(1) View.SYSTEM_UI_FLAG_VISIBLE:系统状态栏和导航栏都显示。

(2) View.SYSTEM_UI_FLAG_FULLSCREEN:显示界面变为全屏,此时不显示系统状态栏。

(3) SYSTEM_UI_FLAG_HIDE_NAVIGATION:隐藏导航栏。

(4) View.SYSTEM_UI_FLAG_IMMERSIVE:设置为沉浸模式。

以上常量可以多个同时使用,中间用竖线"|"分隔。第 48 行如果去掉 View.SYSTEM_UI_FLAG_IMMERSIVE,单击按钮隐藏系统状态栏和导航栏后再次单击屏幕,系统状态栏和导航栏又会再次显示。加入沉浸模式可以避免此类问题。

对于 API 30 的 Android 系统,引入了新的对象和命令,下列代码实现隐藏系统状态栏和导航栏。如果将 hide() 方法换成 show() 方法,则变为显示相关对象。

```
WindowInsetsController windowInsetsController = getWindow().getInsetsController();
    if (windowInsetsController != null)
    {
        if (button3.getText().toString().equals(getString(R.string.API30Visible)))
        {
            windowInsetsController.hide(WindowInsets.Type.statusBars());
            windowInsetsController.hide(WindowInsets.Type.navigationBars());
            button3.setText(getString(R.string.API30Invisible));
        }
    }
```

5.2.2 关闭 Activity

可以通过关闭 Activity 来切换 Activity 或者关闭整个 App。本案例在布局文件中添加了一个 EditText 和 Button。当运行不同的关闭代码时,可以通过列表键对比退出效果。

```
【FirstActivity.java】
01  public class FirstActivity extends Activity
02  {
03      @Override
04      public void onCreate(Bundle savedInstanceState)
05      {
06          super.onCreate(savedInstanceState);
```

```
07                    setContentView(R.layout.main);
08
09                    Button button1 = (Button) findViewById(R.id.button1);
10                    button1.setOnClickListener(new View.OnClickListener()
11                    {
12                        @Override
13                        public void onClick(View v)
14                        {
15                            //方法一
16                            finish();
17                            //方法二
18                            //onDestroy();//注：高版本 SDK 无法关闭 Activity
19                            //方法三
20                            //System.exit(0);
21                            //方法四
22                            //Intent intent = new Intent(Intent.ACTION_MAIN);
23                            //intent.addCategory(Intent.CATEGORY_HOME);
24                            //startActivity(intent);
25                            //方法五 kill 当前进程
26                            //android.os.Process.killProcess(android.os.Process.myPid());
27                        }
28                    });
29                }
30   }
```

每次运行程序时先修改文本输入框中的内容,然后再单击"退出"按钮,执行相应的关闭 Activity 命令。为了方便后续的讲解,术语"执行唤回"是指单击列表键,再单击程序列表中的 HelloAndroid 程序。修改文本输入框中内容的目的是对比执行唤回操作后文本输入框是否能保留关闭前的值。

执行第 16 行的 finish() 方法,单击列表键能看到 HelloAndroid 程序中的控件,执行唤回操作,文本输入框中的值还原为初始值。

第 18 行的 onDestroy() 方法在新版本中已无法关闭 Activity。

执行第 20 行 System.exit(0) 代码退出后,单击列表键在运行程序列表中看不到控件,说明相关资源已被清除。执行唤回操作相当于重启 HelloAndroid 程序。

第 22 行开始的方法四是将系统桌面作为切换的 intent 来处理,等效于按 Home 键。所以程序列表中能看到控件,执行唤回操作后文本输入框中的内容也保持不变。相当于将后台的 HelloAndroid 程序切换回前台。

第 26 行的命令是在 Android 操作系统级执行杀死当前进程操作,因此退出后程序列表中看不到控件。

5.2.3 生命周期

在实际的 App 开发中,Activity 往往不止一个。在多个 Activity 间切换时,存在相关控件是否还保留 Activity 切换前的值,或者是否需要刷新以获取新值的问题。在之前的案例中,控件的初始化赋值等代码都是放在 onCreate() 方法中的,而为了保证 Activity 切换以后程序能按正常逻辑运行,有些代码就需要放在 Activity 的其他方法中。Activity 生命周期如图 5-8 所示。

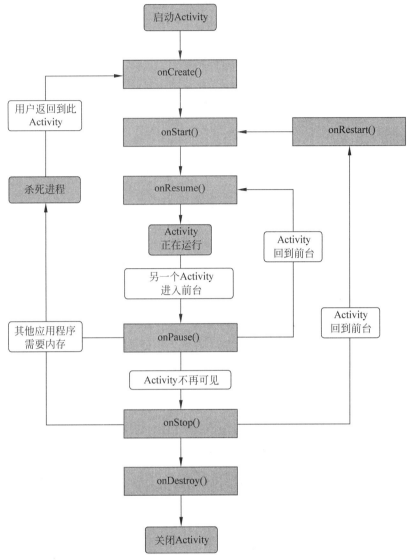

图 5-8 Activity 生命周期

从图 5-8 可以看出,除了第一次启动 Activity 或者已启动的 Activity 资源被 Android 系统释放后重启会执行 onCreate()方法外,其他如 Activity 被遮挡后重新回到前台显示、切换程序后 Activity 回到前台显示等情况都会跳过 onCreate()方法,此时可能从 onRestart() 方法、onStart()方法或 onResume()方法开始执行。因此可根据实际流程,将相关代码放在不同的方法中。为了便于理解,可以把 Activity 分成 3 个不同范围的生命周期。

(1) 完整生命周期:从 onCreate()方法到 onDestroy()方法。

(2) 可见生命周期:从 onStart()方法到 onStop()方法。顾名思义,是 Activity 从显现到消失的周期。如果当前 Activity 被其他 View 部分遮挡,当前 Activity 就还在可见生命周期内;如果是完全遮挡就退出可见生命周期。

(3) 前台生命周期:从 onResume()方法到 onPause()方法,此周期内的控件可见且可交互。只要当前 Activity 被其他 View 遮挡(不论是全部或部分遮挡)都退出前台生命周期。

为了验证 Activity 在生命周期的流转过程,案例中设计两个 Activity,因篇幅限制,本书只列出了 FirstActivity.java 的代码,SecondActivity.java 代码与其大同小异。程序重写了所有生命周期中的方法,两个 Activity 代码不同的地方是各个方法中使用 Log.i()方法输出的内容,指明分别是由哪一个 Activity 的哪一个方法执行的。

【FirstActivity.java】
```
01  public class FirstActivity extends Activity
02  {
03      private Button button1;
04
05      @Override
06      public void onCreate(Bundle savedInstanceState)
07      {
08          Log.i("xj", "第一个 Activity ---> onCreate");
09          super.onCreate(savedInstanceState);
10          setContentView(R.layout.main);
11          button1 = (Button) findViewById(R.id.myButton);
12          button1.setOnClickListener(new ButtonOnClickListener()); //设置监听器
13      }
14
15      @Override
16      protected void onStart()
17      {
18          Log.i("xj", "第一个 Activity ---> onStart");
19          super.onStart();
20      }
21
22      @Override
23      protected void onRestart()
24      {
25          Log.i("xj", "第一个 Activity ---> onRestart");
26          super.onRestart();
27      }
28
29      @Override
30      protected void onResume()
31      {
32          Log.i("xj", "第一个 Activity ---> onResume");
33          super.onResume();
34      }
35
36      @Override
37      protected void onPause()
38      {
```

```
39            Log.i("xj", "第一个 Activity ---> onPause");
40            super.onPause();
41        }
42
43        @Override
44        protected void onStop()
45        {
46            Log.i("xj", "第一个 Activity ---> onStop");
47            super.onStop();
48        }
49
50        @Override
51        protected void onDestroy()
52        {
53            Log.i("xj", "第一个 Activity ---> onDestory");
54            super.onDestroy();
55        }
56
57    //以下方法不属于生命周期触发事件=========================================
58        @Override
59        protected void onSaveInstanceState(Bundle outState)
60        {
61            super.onSaveInstanceState(outState);
62            Log.i("xj", "第一个 Activity ---> onSaveInstanceState");
63        }
64
65        @Override
66        public void onBackPressed()
67        {
68            super.onBackPressed();
69            Log.i("xj", "第一个 Activity ---> onBackPressed()");
70        }
71
72        class ButtonOnClickListener implements OnClickListener
73        {
74            @Override
75            public void onClick(View v)
76            {
77                Intent intent = new Intent();
78                intent.setClass(FirstActivity.this, SecondActivity.class);
79                FirstActivity.this.startActivity(intent);  //启动第二个 Activity
80                //finish();                                //注意观察有无此命令时的变化
81                //System.exit(0);                          //注意观察有无此命令时的变化
82            }
83        }
84    }
```

第58~70行的方法不属于生命周期的方法，onSaveInstanceState()方法主要在切换Activity、单击HOME键、开关电源键等情况时被调用。onBackPressed()方法在单击返回键时被调用。

结合生命周期图,变更以下条件并切换 Activity,观察调试信息的输出变化。

(1) 观察两个 Activity 切换时的状态变化。
(2) 观察单击 HOME 键并再次运行 App 的变化。
(3) 观察横竖屏切换的状态变化(新版本无变化)。
(4) 观察单击列表键后的变化。
(5) 观察单击返回键后的变化。
(6) 观察将 SecondActivity 变为 Dialog 主题后的变化(部分遮挡),可细分为单击按钮返回 FirstActivity、单击 FirstActivity 区域返回和单击返回键 3 种情况。
(7) 观察添加 finish()命令时的变化。
(8) 观察添加 System.exit(0)命令时的变化。
(9) 观察开关电源键时的变化。

5.3 电话及动态授权

Android 系统已经内置完善的拨打电话功能,并提供了相应的调用接口。本案例演示使用系统界面拨号和拨打电话功能。通过拨打电话功能还可以了解 Android 的动态授权的工作机制。

【FirstActivity.java】
```
01   public class FirstActivity extends Activity
02   {
03       @Override
04       public void onCreate(Bundle savedInstanceState)
05       {
06           super.onCreate(savedInstanceState);
07           setContentView(R.layout.main);
08
09           Button button1 = (Button) findViewById(R.id.button1);
10           Button button2 = (Button) findViewById(R.id.button2);
11           Uri uri = Uri.parse("tel:5556");          //设置电话号码
12           //直接调用拨号界面,但并不拨出
13           button1.setOnClickListener(new View.OnClickListener()
14           {
15               @Override
16               public void onClick(View v)
17               {
18                   Intent intent = new Intent(Intent.ACTION_DIAL, uri);
19                   startActivity(intent);
20               }
21           });
22
23           //直接调用拨号界面并拨出
24           button2.setOnClickListener(new View.OnClickListener()
25           {
```

```
26              @Override
27              public void onClick(View v)
28              {
29                  if ( checkSelfPermission ( Manifest. permission. CALL _ PHONE ) ! =
                    PackageManager.PERMISSION_GRANTED)
30                  {
31                      requestPermissions(new String[]{Manifest.permission.CALL_
                        PHONE}, 1);
32                      return;
33                  }
34                  Intent intent = new Intent(Intent.ACTION_CALL, uri);
35                  startActivity(intent);
36              }
37          });
38      }
39      @Override
40      public void onRequestPermissionsResult(int requestCode, @NonNull String[] permissions,
        @NonNull int[] grantResults)
41      {
42          Log.i("xj", "permissions[0] = " + permissions[0] + "\ngrantResults[0] = "
            + grantResults[0]);
43      }
44  }
```

第 11 行定义 Uri 设置要拨打的电话号码，短信、邮件、网址等也可以采用此方式定义。tel 代表电话，5556 是要拨打电话的号码。当前模拟器电话号码是 5554，同时也是模拟器与 adb 服务通信的 TCP 端口号。模拟器完整的电话号码是 15555215554。5555 是 adb 服务的 TCP 端口号。如果启动多个模拟器，第二个模拟器的号码就是 5556，以此类推。本案例为了验证拨出电话功能，需要启动两个模拟器。在 Terminal 窗口中输入 **adb devices** 命令，返回如下已连接的模拟器和真实 Android 设备信息：

```
List of devices attached
9YE0218418013349         device
emulator-5554   device
emulator-5556   device
```

其中，第 1 行的设备信息是真实的手机信息，后面两行是模拟器的信息，模拟器信息中的数字就是电话号码，也是 TCP 端口号。早期版本的模拟器会将此号码直接显示在模拟器窗口的标题栏上。

第 13~21 行定义 button1 单击监听器，完成拨号功能。其中，第 18 行使用 Intent 的 ACTION_DIAL 常量调用 Android 系统的电话拨号界面，并预置 uri 中的电话号码 5556。执行第 19 行启动 intent 命令显示程序初始界面，如图 5-9 所示。

单击 button1（显示"电话拨号，并不拨出"）按钮显示电话拨号界面，如图 5-10 所示。此时要拨打的电话号码 5556 已显示，用户还需要单击拨出键才能完成拨打电话。

图 5-9　程序初始界面　　　　图 5-10　电话拨号界面

第 24～37 行定义 button2 单击监听器,完成拨打电话功能。拨打电话需要申请电话呼叫权限。早期的 Android 版本申请权限只要在 AndroidManifest.xml 中添加 **< uses-permission android:name="android.permission.CALL_PHONE" />** 权限即可。当用户安装 App 时,会显示需要电话呼叫权限界面,但绝大部分用户都不会查看权限列表,而是直接单击"下一步"按钮继续安装,这将导致一些恶意软件乘虚而入。从 API 23 开始,对于一些关键权限(如电话呼叫、发送短信、位置信息、存储读取等)增加了动态授权功能,即用户在运行 App 并使用到相关功能时弹出权限申请界面,如图 5-11 所示。如果选择拒绝,则下一次拨打电话时又会弹出动态授权界面,同时选项中多了一项"拒绝,不要再询问",如图 5-12 所示。如果选择"拒绝,不要再询问"选项,后续再单击拨打电话按钮时不再弹出动态授权界面,

图 5-11　首次提示电话拨打动态授权　　　　图 5-12　多次提示电话拨打动态授权

也就无法完成拨打电话功能。如果选择"允许"选项,则返回程序界面,当再次单击"拨打电话"按钮时,自动调用拨号界面并完成电话拨号和拨打功能,接收方手机将显示来电提示。第 29 行用来判断是否已经授予 CALL_PHONE 权限,如果没有相应权限就执行第 31 行弹出动态权限申请界面。如果已经有权限,则直接执行第 34 行和第 35 行调用系统拨号界面并拨打电话。

"动态授权"对话框与案例程序是异步执行的。本案例中执行第 31 行将弹出"动态授权"对话框,此时用户不进行任何操作,程序也会继续执行。因此在第 32 行加入 return 命令,防止因为未获取拨打电话权限而继续执行第 34~35 行引发异常。也可以将第 34~35 行代码放入 try-catch 中捕获异常,保证程序正常运行。还有一种方式是将第 34~35 行代码放入第 40~43 行的回调方法 onRequestPermissionsResult( )中。当调用第 31 行的 requestPermissions()方法时,系统都会自动调用 onRequestPermissionsResult( )方法返回动态授权的用户操作结果,其中字符串数组 permissions 是动态授权申请的权限,数组长度由动态授权申请的权限数量决定,整型数组 grantResults 是动态授权用户操作结果,−1 代表用户选择了"拒绝"选项,0 代表用户选择了"允许"选项。

5.4 发送短信

下面的案例演示用两种方法发送短信。

```
【FirstActivity.java】
01    public class FirstActivity extends Activity
02    {
03        @Override
04        public void onCreate(Bundle savedInstanceState)
05        {
06            super.onCreate(savedInstanceState);
07            setContentView(R.layout.main);
08
09            Button button1 = (Button) findViewById(R.id.button1);
10            Button button2 = (Button) findViewById(R.id.button2);
11            EditText editText1 = (EditText) findViewById(R.id.editText1);
12            EditText editText2 = (EditText) findViewById(R.id.editText2);
13            editText1.setText("5554");
14            editText2.setText("张三又来了");
15
16            button1.setOnClickListener(new View.OnClickListener()
17            {
18                @Override
19                public void onClick(View v)
20                {
21                    if ( checkSelfPermission ( Manifest. permission. SEND _ SMS ) ! =
                        PackageManager.PERMISSION_GRANTED)
22                    {
23                        requestPermissions(new String[]{Manifest.permission.SEND_
                            SMS}, 1);
```

```
24                  } else
25                  {
26                      //方法一：使用 SmsManager
27                      SmsManager sms = SmsManager.getDefault();
28                      PendingIntent pendingIntent = PendingIntent.getBroadcast
                         (FirstActivity.this, 0, new Intent(), 0);
29                       sms.sendTextMessage(editText1.getText().toString(), null,
                         editText2.getText().toString(), pendingIntent, null);
30                  }
31              }
32          });
33
34          button2.setOnClickListener(new View.OnClickListener()
35          {
36              @Override
37              public void onClick(View v)
38              {
39                  //方法二：使用 Intent 调用系统短信
40                  Uri uri = Uri.parse("smsto://" + editText1.getText().toString());
41                  Intent intent = new Intent(Intent.ACTION_SENDTO, uri);
42                  intent.putExtra("sms_body", editText2.getText().toString());
43                  startActivity(intent);
44              }
45          });
46      }
47  }
```

方法一是使用 SmsManager 发送短信。第 21 行和第 23 行完成动态授权判断和申请功能。第 27～28 行初始化发送短信对象，第 29 行用 sendTextMessage()发出短信。

方法二是使用 Android 系统自带发送短信界面，第 40～43 行的代码只是设置接收短信号码和发送短信的内容，短信还未发送，所以不需要动态授权。

使用 SmsManager 发送短信的代码存在一个问题，每个短信只能容纳 1120b，按每个 ASCII 码 7b 计算，有 160 个字符，折合 GBK 汉字有 70 个字符。如果 ASCII 和汉字混用则全部按汉字计算，总字符也不能超过 70 个。包含汉字的短信如果长度超出 70 个汉字，一种解决方案是在程序中将超出长度的短信进行拆分，然后依次发送。此时接收端会收到多条独立的短信。在实际生活中经常会收到一封短信中包含超过 70 个汉字的内容，这是调用 sendMultipartTextMessage()方法实现的。当长短信超过 70 个汉字时，每条短信的前 48b 作为长短信头部结构标识，所以每条长短信可以发送 67 个汉字。当短信长度小于或等于 70 时，为 1 条短信；当 70<短信长度≤134 时，为 2 条短信，以此类推。接收方按长短信格式将若干条短信合成一条长短信，逻辑上是一条短信，实际上还是由多条短信合成的。使用下述代码即可实现长短信发送。

```
01  SmsManager sms = SmsManager.getDefault();
02  ArrayList<String> smsList = sms.divideMessage(editText2.getText().toString());
03  PendingIntent pendingIntent = PendingIntent.getBroadcast(FirstActivity.this, 0, new
    Intent(), 0);
```

```
04    ArrayList < PendingIntent > sentIntents = new ArrayList < PendingIntent >();
05    for (int i = 0; i < smsList.size(); i++)
06    {
07        sentIntents.add(pendingIntent);
08    }
09    sms.sendMultipartTextMessage ( editText1. getText ( ). toString ( ), null, smsList,
      sentIntents, null);
```

第 2 行调用 divideMessage()方法实现长短信字符串的分割。

第 3~8 行设置长短信相关 Intent。

第 9 行调用 sendMultipartTextMessage()方法发送长短信。

由于发送短信属于重要权限,很多国产手机的 Android 系统会对发送短信权限做出更多提示,有的会弹出风险提示框,提示哪一个应用在试图发送短信,用户选择的权限又分为"本次允许"和"始终允许"等,最终目的都是提高设备使用的安全性。

5.5 Menu

Menu 菜单属于 Android 变动比较大的控件。早期 Android 设备的导航栏按键分别为返回键、主页键和菜单键。如果 Activity 中包含菜单,用户单击菜单键就可以调出菜单。后来 Android 将菜单键改为了列表键,调出菜单的方式也改成了在标题栏上单击菜单图标 。菜单的外观也由早期的在屏幕下方显示的 2×3 列表变成了右上方弹出单列列表。

视频讲解

5.5.1 构建菜单

```
【FirstActivity.java】
01    public class FirstActivity extends Activity
02    {
03        @Override
04        public void onCreate(Bundle savedInstanceState)
05        {
06            super.onCreate(savedInstanceState);
07            setContentView(R.layout.main);
08        }
09
10        @Override
11        public boolean onCreateOptionsMenu(Menu menu)
12        {
13            menu.add(0, 1, 2, "红色"); //添加菜单(int groupId,int itemId,int order)
14            menu.add(0, 2, 2, "绿色");
15
16            menu.add(1, 1, 6, "蓝色");
17            menu.add(1, 2, 5, "紫色");
18
19            menu.add("白色");
20            menu.add("黑色");
```

```
21
22                menu.add(0, 3, 4, "橙色1");
23                menu.add(0, 4, 5, "橙色2");
24                //menu.removeGroup(0); //移除指定组的菜单
25                //menu.setGroupEnabled(0, false);
26                //menu.setGroupVisible(0, false);
27                return super.onCreateOptionsMenu(menu);
28          }
29     }
```

Activity 的菜单是通过第 11~28 行的 onCreateOptionsMenu()方法来构建的,对其自带的形参 menu 执行 add()方法建立菜单项。第 13 行的 add()方法的参数分别为 groupId(菜单项分组号)、itemId(菜单项 id)、order(菜单项排列顺序,按升序排序)和 title(菜单项文字)。如果 order 相同则按 add()方法执行的先后顺序排列。

第 19 行的 add()方法只有 title 参数,则 groupId、itemId 和 order 都为默认值 0。建议对添加的菜单项按功能分组,所有菜单项的 itemId 设置不同的值,便于后续编程时简化判断选择的菜单项。

第 24 行的 removeGroup(0)的作用是将 groupId 为 0 的分组菜单移除。这里有个缺陷一直未修复:当移除分组的菜单项 order 大于或等于后续其他分组菜单项的 order 值,或者 order 小于或等于前方其他分组单项的 order 值时,无法用 removeGroup()方法移除。

第 25 行是将 groupId 为 0 的分组菜单项设置为不可用,菜单文字变为灰色。此时菜单项无法单击使用。

第 26 行是将 groupId 为 0 的分组菜单项设置为不可见。此时菜单项还是分配了地址空间,这与 removeGroup()删除分配空间是不同的。

setGroupVisible(0, false)运行结果如图 5-13 所示。

图 5-13 setGroupVisible(0,false)
运行结果

5.5.2 响应菜单项单击

```
【FirstActivity.java】
01   public class FirstActivity extends Activity
02   {
03        TextView textView1;
04        @Override
05        public void onCreate(Bundle savedInstanceState)
06        {
07             super.onCreate(savedInstanceState);
08             setContentView(R.layout.main);
09             textView1 = (TextView) findViewById(R.id.textView1);
10             textView1.append("\n 执行了 onCreate()");
```

```
11      }
12
13      @Override
14      public boolean onCreateOptionsMenu(Menu menu)
15      {
16          MenuItem menu1 = menu.add(0, 1, 1, "红色");
17          MenuItem menu2 = menu.add(0, 2, 2, "黑色");
18          textView1.append("\n 执行了 onCreateOptionsMenu()")
19          //方法一: 使用监听器响应单击菜单操作,其优先级高于 onOptionsItemSelected()方法
20          MenuItem.OnMenuItemClickListener onMenuItemClickListener = new MenuItem.OnMenuItemClickListener()
21          {
22              @Override
23              public boolean onMenuItemClick(MenuItem item)
24              {
25                  switch (item.getItemId())
26                  {
27                      case 1:
28                          textView1.setTextColor(Color.RED);
29                          textView1.append("\nmenu1.setOnMenuItemClickListener:" +
                                item.getTitle());
30                          break;
31                      case 2:
32                          textView1.setTextColor(Color.BLACK);
33                          textView1.append("\nmenu2.setOnMenuItemClickListener:" +
                                item.getTitle());
34                          break;
35                  }
36                  //return true;   //不再响应 onOptionsItemSelected()方法
37                  return false;
38              }
39          };
40
41          menu1.setOnMenuItemClickListener(onMenuItemClickListener);
42          menu2.setOnMenuItemClickListener(onMenuItemClickListener);
43          return super.onCreateOptionsMenu(menu);
44      }
45
46      //方法二: 内置 onOptionsItemSelected()方法
47      @Override
48      public boolean onOptionsItemSelected(MenuItem item)
49      {
50          switch (item.getItemId())
51          {
52              case 1:
53                  textView1.setTextColor(Color.RED);
54                  break;
55              case 2:
56                  textView1.setTextColor(Color.BLACK);
57                  break;
```

```
58              }
59              textView1.append("\n菜单组" + item.getGroupId() + "\n菜单项" + item.
                getItemId() + "\n菜单顺序:" + item.getOrder() + "\n菜单标题:" + item.
                getTitle());
60              return super.onOptionsItemSelected(item);
61          }
62      }
```

当用户单击菜单项时,响应单击操作的方法有如下两种。

方法一:使用OnMenuItemClickListener监听器响应菜单操作,其优先级高于Activity下的onOptionsItemSelected()方法。第25行通过getItemId()方法获取菜单项的itemId,如果itemId是唯一的,就无须再去判断分组或者是菜单项的title,可以简化判断选中菜单项的代码。监听器的返回值如果采用第36行的true,将不再执行onOptionsItemSelected()方法。监听器返回true的Menu运行结果如图5-14所示。如果设为false,在执行完监听器以后接着执行onOptionsItemSelected()方法。监听器返回false的Menu运行结果如图5-15所示。

图5-14 监听器返回true的Menu运行结果

图5-15 监听器返回false的Menu运行结果

方法二:使用Activity下的onOptionsItemSelected()方法。方法的框架可依次选择Android Studio菜单Code→Override Methods,在弹出的窗口中选择onOptionsItemSelected()方法,Android Studio会构建此方法的代码框架。onOptionsItemSelected()方法的使用与方法一中的onCreateOptionsMenu()方法类似,两者的关系:onCreateOptionsMenu()方法在Activity启动时运行一次,可初始化菜单和菜单项单击监听器,用户单击菜单项由菜单项单击监听器响应,如果监听器的返回值为false,则继续调用onOptionsItemSelected()方法。

用户在使用App的过程中可能会无暇关注右上角的菜单按钮,有两种解决方案。

(1) 在按钮的监听器中加入"**openOptionsMenu();**"命令,当用户单击按钮时执行此命令,自动运行onCreateOptionsMenu()方法实现弹出菜单。

(2) 长按某个控件弹出上下文菜单ContextMenu。

5.5.3 ContextMenu

用户可长按不同的控件弹出不同的上下文菜单ContextMenu,前提条件是这些控件都注册了ContextMenu。

视频讲解

【FirstActivity.java】
```
01  public class FirstActivity extends Activity
02  {
03      TextView textView1, textView2;
```

```
04      EditText editText1;
05
06      @Override
07      public void onCreate(Bundle savedInstanceState)
08      {
09          super.onCreate(savedInstanceState);
10          setContentView(R.layout.main);
11          textView1 = (TextView) findViewById(R.id.textView1);
12          textView2 = (TextView) findViewById(R.id.textView2);
13          editText1 = (EditText) findViewById(R.id.editText1);
14
15          registerForContextMenu(textView1);  //注册 textView1 上下文菜单
16          registerForContextMenu(editText1);  //注册 editText1 上下文菜单
17      }
18
19      @Override
20      public void onCreateContextMenu(ContextMenu menu, View v, ContextMenuInfo menuInfo)
21      {
22          textView2.append("\n重新初始化 menu");
23          if (v == textView1)                        //如果长按 TextView
24          {
25              menu.add(0, 1, 1, "红色");
26              menu.add(0, 2, 2, "黑色");
27              menu.setHeaderTitle("请选择 TextView 字体颜色");  //设置标题栏文字
28              return;
29          }
30          //如果长按 EditText
31          menu.add(1, 3, 11, "红色");
32          menu.add(1, 4, 12, "蓝色");
33          menu.setHeaderTitle("请选择 EditText 字体颜色");
34          menu.setHeaderIcon(R.drawable.icon);  //设置标题栏图标
35
36          super.onCreateContextMenu(menu, v, menuInfo);
37      }
38
39      @Override
40      public boolean onContextItemSelected(MenuItem item)
41      {
42          switch (item.getItemId())              //判断被单击的菜单项
43          {
44              case 1:
45                  textView1.setTextColor(Color.RED);
46                  break;
47              case 2:
48                  textView1.setTextColor(Color.BLACK);
49                  break;
50              case 3:
51                  editText1.setTextColor(Color.RED);
52                  break;
53              case 4:
```

```
54                    editText1.setTextColor(Color.BLUE);
55                    break;
56            }
57            return super.onContextItemSelected(item);
58      }
59
60      @Override
61      public void onContextMenuClosed(Menu menu)
62      {
63            textView2.append("\n退出了上下文菜单:" + menu.toString());
64            super.onContextMenuClosed(menu);
65      }
66  }
```

第 15～16 行通过 registerForContextMenu()方法将注册 textView1 和 editText1 上下文菜单。长按这两个控件就会自动调用第 20 行的 onCreateContextMenu()方法。

第 22 行的代码是为了演示每次调用上下文菜单时,都会重新运行 onCreateContextMenu()方法,这与 Menu 的 onCreateOptionsMenu()方法只运行一次是不同的。

第 61 行的 onContextMenuClosed()方法在退出上下文菜单时运行,menu.toString()会显示上下文菜单的地址,注意观察输出结果,即使单击相同的菜单项,每一次的上下文菜单地址也是不同的。

第 23～29 行是生成 textView1 的 ContextMenu,第 31～34 行生成 editText1 的 ContextMenu。

第 40～58 行是上下文菜单项被单击后的响应代码,与之前案例的 Menu 代码类似。

【注】 如果没有特别提示,长按控件才弹出上下文菜单的方式很容易被用户忽略。每一次都要重新构建上下文菜单,效率也略显低下。

5.6 Notification

Notification(通知)可独立于 App 显示在系统下拉通知栏中。由于很多 App 会发送大量的通知,Android 对通知的限制也日趋严格。

```
【FirstActivity.java】
01  public class FirstActivity extends Activity
02  {
03      Notification notification;                    //Notification 对象
04      NotificationManager notificationManager;      //NotificationManager 对象
05      static final int NOTIFY_ID = 1234;            //通知管理器通过通知 id 来标识不同的通
                                                      //知,创建和删除通知需要提供通知 id
06
07      @Override
08      public void onCreate(Bundle savedInstanceState)
09      {
10          super.onCreate(savedInstanceState);
11          setContentView(R.layout.main);
```

```java
12
13          Button button1 = (Button) findViewById(R.id.button1);
14          Button button2 = (Button) findViewById(R.id.button2);
15          TextView textView1 = (TextView) findViewById(R.id.textView1);
16
17          button1.setOnClickListener(new View.OnClickListener()
18          {
19              @Override
20              public void onClick(View v)
21              {
22                  //NotificationManager 是一个系统 Service,必须通过
                    //getSystemService()方法来获取
23                  notificationManager = (NotificationManager) getSystemService(NOT-
                    IFICATION_SERVICE);
24
25                  //API 26 以上需要建立 CHANNEL
26                  if (android.os.Build.VERSION.SDK_INT >= android.os.Build.VERSION_
                    CODES.O)
27                  {
28                      String Channel_Id = "1";                    //定义通道 id
29                      CharSequence name = "my_channel";           //通道名称
30                      int importance = NotificationManager.IMPORTANCE_HIGH;
                                                                    //通知的重要级别
31
32                      NotificationChannel notificationChannel = new NotificationChannel
                        (Channel_Id, name, importance);
33                      notificationChannel.setDescription("Description");
34                      notificationChannel.enableLights(true);
35                      notificationChannel.setLightColor(Color.RED);
36                      notificationChannel.enableVibration(true);   //允许振动
37                      notificationChannel.setVibrationPattern(new long[]{100, 200, 300,
                        400, 500, 400, 300, 200, 400});
38                      notificationManager.createNotificationChannel(notificationChannel);
39
40                      Notification.Builder builder = new Notification.Builder
                        (getApplicationContext(), Channel_Id)
41                              .setSmallIcon(R.drawable.icon)
                        //设置状态栏中的小图片,尺寸一般建议为 24×24,这
                        //个图片同样也是在下拉状态栏中所显示,如果在那里
                        //需要更换更大的图片,可以使用 setLargeIcon
42                              .setContentTitle("Notification 标题")
43                              .setContentText("Notification 内容");
44
45                      Intent resultIntent = new Intent(getApplicationContext(),
                        FirstActivity.class);
46                      TaskStackBuilder stackBuilder = TaskStackBuilder.create(get-
                        ApplicationContext());
47                      stackBuilder.addParentStack(FirstActivity.class);
48                      stackBuilder.addNextIntent(resultIntent);
```

```java
49                    PendingIntent resultPendingIntent = stackBuilder.getPendingIntent
                          (0, PendingIntent.FLAG_UPDATE_CURRENT);
50                    builder.setContentIntent(resultPendingIntent);
51                    builder.setAutoCancel(true);//单击通知就自动消除.必须与
                                             //setContentIntent 一起使用才有效
52
53                    notificationManager.notify(NOTIFY_ID, builder.build());
                                             //发出通知
54                }
55                else
56                {
57                    //API 26 版本以下
58                    Intent intent = new Intent(getApplicationContext(),FirstAc-
                          tivity.class);
59                    PendingIntent pendingIntent = PendingIntent.getActivity(get-
                          ApplicationContext(), 0, intent, 0);
60                    notification = new Notification.Builder(FirstActivity.this)
61                            .setSmallIcon(R.drawable.icon)
62                            .setTicker("Notification 测试")
63                            .setContentTitle("老版本 Notification 标题")
64                            .setContentText("老版本 Notification 内容")
65                            .setVibrate(new long[] { 500L, 200L, 200L, 500L })
66                            .setContentIntent(pendingIntent)
67                            .build();
68                    notificationManager.notify(NOTIFY_ID, notification);
69                }
70            }
71        });
72
73        button2.setOnClickListener(new View.OnClickListener()
74        {
75            @Override
76            public void onClick(View v)
77            {
78                try
79                {
80                    notificationManager.cancel(NOTIFY_ID);
                      //取消通知.如果相关通知已经手工清除,再次调用此命令会出错
81                    //notificationManager.cancelAll();   //清除所有通知
82                }
83                catch (Exception e)
84                {
85                    textView1.append(e.toString());
86                }
87            }
88        });
89    }
90 }
```

从 API 26 起,使用 NotificationManager 时需要加入通道号。程序提供了新老两个版本发送通知的代码,集合了灯光、振动等功能。Notification 可视为一个容器,显示的文字是在其中的 TextView 中,还可以添加图片甚至进度条。很多音乐播放软件设计了在通知栏中显示简单的播放界面。

第 23 行定义了 NotificationManager,用于发送通知。

第 26~54 行是 API 26 及以上版本发送通知代码,老版本的代码在第 56~69 行。

第 51 行一般设置为 true,即用户单击通知栏中的通知就自动删除该通知,也可以单击"取消通知"按钮清除通知。如果设置为 false 或者注释此行,通知栏中通知需要执行第 80 行或第 81 行代码才能删除。如果通知已经手工删除,再次单击"取消通知"按钮会抛出异常。

5.7 Service

Service(服务)是在后台可长时间运行的组件。Service 有如下两种状态。

(1) Started:此状态是 Service 通过 startService()启动服务,且可保持 Service 一直运行。其 Service 生命周期如图 5-16 的左边流程所示。

(2) Bounded:Service 是通过 bindService()绑定了服务,此时允许组件与 Service 进行"请求-应答"方式的数据交互。其 Service 生命周期如图 5-16 的右边流程。

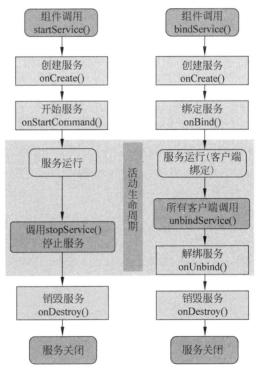

图 5-16　Service 生命周期

将 Service 生命周期中对应方法放入 MyService.java 中。

【MyService.java】
```
01  public class MyService extends Service
02  {
03      /**
04       * 在 MyBinder 直接继承 Binder 而不是 IBinder,因为 Binder 实现了 IBinder 接口,使用更方便。
05       */
06      public class MyBinder extends Binder
07      {
08          //定义加法
09          public int add(int a, int b)
10          {
11              return a + b;
12          }
13      }
14
15      public MyBinder myBinder;                    //定义公共成员变量
16
17      @Override
18      public void onCreate()
19      {
20          myBinder = new MyBinder();     //生成实例
21          com.xiaj.FirstActivity.textView1.append("Service 的 onCreate 方法\n");
22          super.onCreate();
23      }
24
25      @Override
26      public IBinder onBind(Intent arg0)
27      {
28          com.xiaj.FirstActivity.textView1.append("Service 的 onBind 方法\n");
29          return myBinder;              //必须返回实例才能激活 ServiceConnection 中的回调
                                          //方法 onServiceConnected()
30      }
31
32      @Override
33      public boolean onUnbind(Intent intent)
34      {
35          com.xiaj.FirstActivity.textView1.append("Service 的 onUnbind()方法\n");
36          return super.onUnbind(intent);
37      }
38
39      @Override
40      public void onDestroy()
41      {
42          com.xiaj.FirstActivity.textView1.append("Service 的 onDestroy()方法\n");
43          super.onDestroy();
44      }
45
46      @Override
47      public int onStartCommand(Intent intent, int flags, int startId)
```

```
48    {
49        com.xiaj.FirstActivity.textView1.append("Service 的 onStartCommand()方法\n");
50        return super.onStartCommand(intent, flags, startId);
51    }
52
53    @Override
54    public void onRebind(Intent intent)
55    {
56        com.xiaj.FirstActivity.textView1.append("Service 的 onRebind()方法\n");
57        super.onRebind(intent);
58    }
59 }
```

第 6~13 行定义了继承 Binder 的 MyBinder 类,内部定义了 add()方法实现两个整数相加。其他方法都对应生命周期中的方法。

第 21 行在 Service 中要对 FirstActivity 中控件操作需加 Activity 前缀,如果有同名 Activity 还需加 package 前缀。完整名称如下:

package名称 + Activity名称 + 控件名称

Service 要在 AndroidManifest.xml 文件中注册才能使用。

【AndroidManifest.xml】
```
01  <?xml version = "1.0" encoding = "UTF - 8"?>
02  < manifest xmlns:android = "http://schemas.android.com/apk/res/android"
03      package = "com.xiaj">
04
05      < application
06          android:icon = "@drawable/icon"
07          android:label = "@string/app_name">
08          < activity
09              android:name = "com.xiaj.FirstActivity"
10              android:label = "@string/app_name">
11              < intent - filter >
12                  < action android:name = "android.intent.action.MAIN" />
13                  < category android:name = "android.intent.category.LAUNCHER" />
14              </intent - filter >
15          </activity >
16          < service android:name = "com.xiaj.MyService">
17              < intent - filter >
18                  < action android:name = "com.xiaj.MY_SERVICE" />
19              </intent - filter >
20          </service >
21
22      </application >
23  </manifest >
```

【FirstActivity.java】
```
01  public class FirstActivity extends Activity
02  {
03    public static TextView  textView1;
04
05    @Override
06    public void onCreate(Bundle savedInstanceState)
07    {
08        super.onCreate(savedInstanceState);
09        setContentView(R.layout.main);
10        Button button1 = (Button) findViewById(R.id.button1);
11        Button button2 = (Button) findViewById(R.id.button2);
12        Button button3 = (Button) findViewById(R.id.button3);
13        Button button4 = (Button) findViewById(R.id.button4);
14        textView1 = (TextView) findViewById(R.id.textView1);
15
16        Intent intent = new Intent(getApplicationContext(), MyService.class);
            //将 MyService 绑定到 intent
17           //intent.setAction(MY_SERVICE); //对新版本,service 的调用必须是显式 intent
                                   //调用,用隐式 intent 调用会出错
18
19        //启动 Service 按钮
20        button1.setOnClickListener(new View.OnClickListener()
21        {
22            @Override
23            public void onClick(View v)
24            {
25                startService(intent); //启动 Service
26                //第一次单击会依次调用 onCreate()和 onStart()方法,第二次单击只调用
                  //onStart()方法
27                //并且系统只会创建 Service 的一个实例(因此停止 Service 只需要一次
                  //stopService()调用)
28            }
29        });
30        //停止 Service 按钮
31        button2.setOnClickListener(new View.OnClickListener()
32        {
33            @Override
34            public void onClick(View v)
35            {
36                stopService(intent); //停止 Service
37            }
38        });
39
40        ServiceConnection serviceConnection = new ServiceConnection()
41        {
42            //连接对象,在正常单击按钮时不会触发以下方法
43            @Override
44            public void onServiceDisconnected(ComponentName name)
45            {
```

```
46              textView1.append("Service 连接断开\n");
47          }
48
49          @Override
50          public void onServiceConnected(ComponentName name, IBinder service)
51          {
52              textView1.append("Service 连接成功\n");
53              MyService.MyBinder myBinder = (MyService.MyBinder)service;
54              textView1.append("调用服务计算: 1 + 2 = " + myBinder.add(1, 2) + "\n");
                //调用服务中的 add()方法
55          }
56      };
57
58      //绑定 Service 按钮
59      button3.setOnClickListener(new View.OnClickListener()
60      {
61          @Override
62          public void onClick(View v)
63          {
64              bindService(intent, serviceConnection, Context.BIND_AUTO_CREATE);
65          }
66      });
67
68      //解除绑定 Service 按钮
69      button4.setOnClickListener(new View.OnClickListener()
70      {
71          @Override
72          public void onClick(View v)
73          {
74              try
75              {
76                  //执行 unbindService()之后会依次调用 onUnbind()和 onDestroy()
77                  unbindService(serviceConnection); //解除绑定 Service
78              }
79              catch(Exception ex)
80              {
81                  textView1.append("Service 未绑定,无法执行 unbindService 解除绑定\n");
82              }
83
84          }
85      });
86  }
87 }
```

单击 button1 按钮运行第 25 行 startService()方法启动 MyService 服务,MyService 执行图 5-16 的 onCreate()和 onStartCommand()方法。

单击 button2 按钮执行第 36 行 stopService()方法停止 MyService 服务,MyService 执行 onDestroy()方法。

单击 button3 执行图 5-16 的 onCreate( )和 onBind( )方法。onBind( )方法中返回 myBinder 对象会激活 FirstActivity.java 的 ServiceConnection 对象的回调方法 onServiceConnected( )（第 50 行），其中的形参 service 对应 MyService 中的 myBinder。在第 54 行 myBinder.add( )方法就是向 Service 调用 MyBinder 类中的 add( )方法，由此完成 FirstActivity 中单击 button3 调用 Service 中的 add( )计算。

单击 button4 执行 onUnbind( )和 onDestroy( )方法。两条线的生命周期是可以交叉的，如单击 button1 再单击 button2，BindService 会跳过已经运行的 onCreate( )方法，直接从 onBind( )方法开始执行。

5.8 Broadcast

Broadcast（广播）分为发送者和接收者，可实现跨应用的消息传递。重启手机、闹钟、来电、接收短信等都会发出广播，通过 BroadcastReceiver 就可以接收广播并进行相应处理。

5.8.1 静态注册

视频讲解

静态注册是指在 AndroidManifest.xml 中注册广播接收器。定义一个 MyReceiver，需要添加如下标记：

```
<receiver android:name = "com.xiaj.MyReceiver">
    <intent-filter>
        <action android:name = "com.xiaj.MY_RECEIVER" />
    </intent-filter>
</receiver>
```

广播接收器定义在 MyReceiver.java 文件中。

【MyReceiver.java】
```
01  public class MyReceiver extends BroadcastReceiver
02  {
03      @Override
04      public void onReceive(Context context, Intent intent)
05      {
06          String message = "我是 MyReceiver,收到的广播为: " + intent.getStringExtra("message");
07          Toast.makeText(context, message, Toast.LENGTH_LONG).show();
08      }
09  }
```

MyReceiver 类中只有一个方法 onReceive( )，当广播接收器收到广播时就运行 onReceive( ) 方法。其中的形参 intent 包含收到的广播消息键值对，通过 getStringExtra( )就可以得到对应的广播消息。

【FirstActivity.java】
```
01  public class FirstActivity extends Activity
02  {
```

```
03      @Override
04      public void onCreate(Bundle savedInstanceState)
05      {
06          super.onCreate(savedInstanceState);
07          setContentView(R.layout.main);
08          Button button1 = (Button) findViewById(R.id.button1);
09          button1.setOnClickListener(new View.OnClickListener()
10          {
11              @Override
12              public void onClick(View v)
13              {
14                  Intent intent = new Intent();
15                  intent.setAction("com.xiaj.MY_RECEIVER"); //通过 Action 查找 MyReceiver
16                  intent.putExtra("message", "开始点名了"); //设置广播的消息
17                  intent.setPackage(getPackageName());      //指定广播接收者的包名(老
                                                              //版本可以不用这条命令)
18                  sendBroadcast(intent);                    //发送广播
19              }
20          });
21      }
22  }
```

第 15 行通过 setAction()方法设置要在 AndroidManifest.xml 文件中查找的接收器。

第 16 行设置要广播的消息键值对。对于早期的版本到此 Intent 设置完毕,可以直接运行第 18 行的 sendBroadcast()方法发送广播。由于发送广播的 App 实在太多,Android 的新版本 SDK 对发送广播做了限制,需要指定广播接收者的包名,第 17 行的 setPackage()方法可完成此项功能。

发送和接收广播运行结果如图 5-17 所示。

图 5-17　发送和接收广播运行结果

5.8.2 动态注册

动态注册广播接收器的使用方式更加灵活，可以不用在 AndroidManifest.xml 中注册，发送广播报文时也不用指定接收者的包名。

视频讲解

```
【FirstActivity.java】
01  public class FirstActivity extends Activity
02  {
03      final String MY_RECEIVER = "com.xiaj.MY_RECEIVER";
04      MyReceiver receiver;  //定义一个自定义的myReceiver类对象
05
06      @Override
07      public void onCreate(Bundle savedInstanceState)
08      {
09          super.onCreate(savedInstanceState);
10          setContentView(R.layout.main);
11          Button button1 = (Button) findViewById(R.id.button1);
12          Button button2 = (Button) findViewById(R.id.button2);
13          Button button3 = (Button) findViewById(R.id.button3);
14          button1.setOnClickListener(new View.OnClickListener()
15          {
16              @Override
17              public void onClick(View v)
18              {
19                  Intent intent = new Intent();              //Intent 对象
20                  intent.setAction(MY_RECEIVER);             //设置 Action
21                  intent.putExtra("message", "开始点名了");    //设置广播的消息
22                  sendBroadcast(intent);                     //发送广播
23              }
24          });
25
26          button2.setOnClickListener(new View.OnClickListener()
27          {
28              @Override
29              public void onClick(View v)
30              {
31                  IntentFilter filter = new IntentFilter(MY_RECEIVER);
32                  receiver = new MyReceiver();  //初始化自定义的 MyReceiver 类
33                  registerReceiver(receiver, filter);//注册广播接收器,可以多次注册,
                                                  //意味着收到一条广播时会有多个
                                                  //Receiver 接收(注意 toast 显示的
                                                  //次数和时间)
34              }
35          });
36
37          button3.setOnClickListener(new View.OnClickListener()
38          {
39              @Override
40              public void onClick(View v)
```

```
41                          {
42                              unregisterReceiver(receiver); //注销广播接收器
43                          }
44                      });
45              }
46      }
```

第 19~22 行封装 Intent 后发送广播,注意代码中不需要静态注册中设定广播接收者的包名"intent.setPackage(getPackageName());"代码。此时单击 button1 发出的广播对于接收器 MyReceiver 是收不到的,因为此时 MyReceiver 既没有静态注册,也没有动态注册。

第 26~35 行 button2 的单击监听器实现广播接收器的动态注册,关键代码是第 33 行的 registerReceiver()方法动态绑定广播接收器。如果连续单击 button2 按钮,将会注册同等数量的广播接收器,此时再单击 button1,发出的一次广播会被多个广播接收器接收,并依次弹出 Toast。

第 37~44 行 button3 的单击监听器实现注销广播接收器的功能。如果连续多次单击 button3 会出错。可对第 42 行代码设置 try-catch 捕获异常,查看出错的原因,这里就不再赘述。

5.8.3 多接收器接收普通广播

视频讲解

本案例定义两个广播接收器,当发出广播时,两个广播接收器都会收到广播报文。发送广播报文的代码与静态注册案例相同,本节就不再重复讲解。以下是定义的两个接收器类。

【FirstReceiver.java】
```
01  public class FirstReceiver extends BroadcastReceiver
02  {
03      @Override
04      public void onReceive(Context context, Intent intent)
05      {
06          String str = "FirstReceiver 接收到的广播为: " + intent.getStringExtra("message");
07          Log.i("xj", "FirstReceiver: " + str);
08      }
09  }
```

【SecondReceiver.java】
```
01  public class SecondReceiver extends BroadcastReceiver
02  {
03      @Override
04      public void onReceive(Context context, Intent intent)
05      {
06          String str = "SecondReceiver 接收到的广播为: " + intent.getStringExtra("message");
07          Log.i("xj", "SecondReceiver: " + str);
08      }
09  }
```

当 FirstActivity 中发送广播后,两个接收器都会收到广播,调用各自接收器中的

onReceive 方法,通过 Log.i() 方法输出如下信息:

```
I: FirstReceiver: FirstReceiver 接收到的广播为:开始点名了
I: SecondReceiver: SecondReceiver 接收到的广播为:开始点名了
```

此时接收广播的顺序与广播接收器的注册先后顺序相关。

5.8.4 有序广播

视频讲解

如果发送的广播是有序广播,多个广播接收器可按指定的优先级依次接收处理报文。如果优先级高的接收器将消息处理后重新发出,后续的接收器可以同时接收处理前和处理后的广播消息。优先级高的广播接收器也可以终止后续广播接收器接收广播。

```
【AndroidManifest.xm】
01  <?xml version = "1.0" encoding = "UTF - 8"?>
02  < manifest xmlns:android = "http://schemas.android.com/apk/res/android"
03      package = "com.xiaj">
04
05      < permission
06          android:name = "myreceiver.permission.MY_BROADCAST_PERMISSION"
07          android:protectionLevel = "normal" />
08      < uses - permission android:name = "myreceiver.permission.MY_BROADCAST_PERMISSION" />
09
10      < application
11          android:icon = "@drawable/icon"
12          android:label = "@string/app_name">
13          < activity
14              android:name = "com.xiaj.FirstActivity"
15              android:label = "@string/app_name">
16              < intent - filter >
17                  < action android:name = "android.intent.action.MAIN" />
18                  < category android:name = "android.intent.category.LAUNCHER" />
19              </intent - filter >
20          </activity>
21
22          < receiver android:name = "com.xiaj.FirstReceiver">
23              < intent - filter android:priority = "999">
24                  < action android:name = "com.xiaj.MY_RECEIVER" />
25              </intent - filter >
26          </receiver>
27          < receiver android:name = "com.xiaj.SecondReceiver">
28              < intent - filter android:priority = "990">
29                  < action android:name = "com.xiaj.MY_RECEIVER" />
30              </intent - filter >
31          </receiver>
32
33      </application>
34  </manifest>
```

第5～7行的 permission 标签声明自定义权限,第8行的 uses-permission 标签申请自定义权限,其中的权限名称对应 FirstActivity.java 中 sendOrderedBroadcast()方法的第二个参数 receiverPermission 的值。如果没有权限则有序广播无效。

第23～26行和第28～31行定义了广播接收器及优先级,数字越大,优先级越高,将优先接收到有序广播。

FirstActivity.java 文件的代码与5.8.3节案例的代码基本相同,只是将发送广播的代码"sendBroadcast(intent);"改为:

```
sendOrderedBroadcast(intent, "myreceiver.permission.MY_BROADCAST_PERMISSION");
```

上述代码发送的广播即为有序广播,下面是两个接收器处理有序广播的代码。

【FirstReceiver.java】
```
01  public class FirstReceiver extends BroadcastReceiver
02  {
03    @Override
04    public void onReceive(Context context, Intent intent)
05    {
06        String str = "接收到的广播消息为: " + intent.getStringExtra("message");
07        Log.i("xj", "FirstReceiver: " + str);
08
09        //将 message 重新处理后发送出去
10        Bundle bundle = new Bundle();
11        bundle.putString("message", "此广播已被 FirstReceiver 处理过.");
12        setResultExtras(bundle);
13
14        //终止低优先级的 Receiver 接收广播,应用场景: 制作短信接收 app,阻止
           //Android 系统再接收短信
15        //abortBroadcast();
16    }
17  }
```

第6行是按正常方式获取广播报文。

第10～11行是重新封装新的广播报文,再通过第12行的 setResultExtras()方法将处理过的广播消息发送出去,加上 FirstActivity 中发送广播,此时就有两条广播报文了。

第15行如果去掉注释,将阻止后续的广播接收器接收广播,即 SecondReceive 将忽略之前发送的两条广播报文。

【SecondReciever.java】
```
01  public class SecondReceiver extends BroadcastReceiver
02  {
03    @Override
04    public void onReceive(Context context, Intent intent)
05    {
06        String str = "接收到的广播消息为:\n已处理消息: " + getResultExtras(true).getString
```

```
07              ("message") + "\n未处理消息: " + intent.getStringExtra("message");
08          //注意: 已处理消息不再是 intent.getStringExtra("message")
09          Log.i("xj", "SecondReceiver: " + str);
10      }
11  }
```

使用 getResultExtras(true).getString("message") 获取处理过的广播，原始的广播报文还是用 getStringExtra() 方法获取。

有序广播程序运行结果如下：

> I: FirstReceiver:接收到的广播消息为: 来自FirstActivity广播的消息!
> I: SecondReceiver:接收到的广播消息为:
>   已处理消息: 此广播已被FirstReceiver处理过.
>   未处理消息: 来自FirstActivity广播的消息!

如果 FirstReceiver.java 中运行"**abortBroadcast()；**"命令，程序运行结果如下：

> I: FirstReceiver:接收到的广播消息为: 来自FirstActivity广播的消息!

可以看到 SecondReceiver 无法接收任何广播了。

5.9 SQLiteDatabase

Android 内置了 SQLite 数据库，可通过 SQLiteDatabase 类对 SQLite 数据库进行增、删、改、查操作。SQLite 数据库属于文件数据库，能满足 Android 设备日常应用的数据管理需求。本案例主要对数据库和数据表的创建及增、删、改、查等基本操作做一个大致的介绍。

【FirstActivity.java】
```
01  public class FirstActivity extends Activity
02  {
03      SQLiteDatabase myDB; //声明数据库
04      EditText editText1, editText2, editText3;
05      TextView textViewShow;
06
07      @Override
08      public void onCreate(Bundle savedInstanceState)
09      {
10          super.onCreate(savedInstanceState);
11          setContentView(R.layout.main);
12
13          Button button1 = (Button) findViewById(R.id.button1);
14          Button button2 = (Button) findViewById(R.id.button2);
15          Button button3 = (Button) findViewById(R.id.button3);
16          editText1 = (EditText) findViewById(R.id.editTextName);
17          editText2 = (EditText) findViewById(R.id.editTextScore);
```

```java
18          textViewShow = (TextView) findViewById(R.id.textViewShow);
19
20          myDB = openOrCreateDatabase("mydb.db", MODE_PRIVATE, null);
            //第一次创建数据库,后续打开数据库
21          //如用openDatabase()则数据库的路径为/data/data/com.xiaj/databases/mydb.db
22          try
23          {
24              myDB.execSQL("CREATE TABLE Score (Id INTEGER PRIMARY KEY, Name VARCHAR
                (30) NOT NULL DEFAULT '' COLLATE NOCASE, Score FLOAT NOT NULL DEFAULT 0,
                InDate DATETIME NOT NULL DEFAULT(DATETIME('now', 'localtime') ) );");
                                                            //执行无返回数据的SQL命令
25          } catch (Exception e)
26          {
27          }
28          myDB.execSQL("delete from Score");         //清空表中数据
29
30          ContentValues cv = new ContentValues();
31          cv.put("Name", "刘备");
32          cv.put("Score", 95);
33          myDB.insert("Score", null, cv);
34
35          //使用execSQL().注:Sqlite中的execSQL()一次只能执行一条SQL语句,解决的
            //方法使用存储过程或事务
36          try
37          {
38              myDB.beginTransaction();
39              myDB.execSQL("insert into Score(Name, Score) values('关羽', 80)");
40              myDB.execSQL("insert into Score(Name, Score) values('张飞', 90)");
41              //设置事务标志为成功,当结束事务时就会提交事务
42              myDB.setTransactionSuccessful();
43          } catch (Exception e)
44          {
45              e.printStackTrace();
46          } finally
47          {
48              myDB.endTransaction();                  //结束事务
49          }
50          showTable();
51          myDB.close();                               //数据库用完一定要关闭
52
53          View.OnClickListener onClickListener = new View.OnClickListener()
54          {
55              @Override
56              public void onClick(View v)
57              {
58                  myDB = openOrCreateDatabase("mydb.db", MODE_PRIVATE, null);
59                  switch (v.getId())
60                  {
61                      case R.id.button1:
62                          //mydb.execSQL("insert into Score(Name, Score) values('" +
                            edit1.getText() + "'" + ", " + edit2.getText() + ")");
```

```java
63                              //Sqlite数字字段时可以输入字符,SQL语句要加入单引号
64                              myDB.execSQL("insert into Score(Name, Score) values('" +
                                    editText1.getText() + "'" + ", '" + editText2.
                                    getText() + "')");
65                              break;
66                          case R.id.button2:
67                              myDB.execSQL("delete from Score where Name = '" +
                                    editText1.getText() + "'");
68                              break;
69                          case R.id.button3:
70                              myDB.execSQL("update Score set Score = '" + editText2.
                                    getText() + "'where Name = '" + editText1.getText() + "'");
71                              break;
72                      }
73                      showTable();
74                      myDB.close();
75                  }
76              };
77          button1.setOnClickListener(onClickListener);
78          button2.setOnClickListener(onClickListener);
79          button3.setOnClickListener(onClickListener);
80      }
81
82      public void showTable()
83      {
84          Cursor cursor = myDB.rawQuery("select * from Score", null);
85          String column = "";
86          String data = "";
87          while (cursor.moveToNext())
88          {
89              for (int i = 0; i < cursor.getColumnCount(); i++)
90              {
91                  if (cursor.isFirst())            //输出列名
92                  {
93                      column += cursor.getColumnName(i);
94                      if (i != cursor.getColumnCount() - 1)
95                      {
96                          column += "\t\t\t";
97                      }
98                  }
99                  data += cursor.getString(i);   //输出数据
100                 if (i != cursor.getColumnCount() - 1)
101                 {
102                     data += "\t\t\t\t";
103                 }
104             }
105             data += "\n";
106         }
107         textViewShow.setText(column + "\n……………………
                ………………\n" + data + "\n共计:" + cursor.getCount() + "行,
                " + cursor.getColumnCount() + "列");
```

```
108            cursor.close();
109        }
110    }
```

SQLite 数据库运行结果如图 5-18 所示。

第 3 行声明 SQLiteDatabase 变量 myDB，供后续对数据库进行增、删、改、查操作。

第 20 行调用 openOrCreateDatabase() 方法建立或打开数据库 mydb.db。当第一次运行此命令时建立数据库，当数据库已经建立后再运行此命令则变为打开数据库。新建立的数据库所在目录是 /data/data/com.xiaj/databases。当卸载 App 时，数据库和相关目录也会自动删除。

第 22～27 行是调用 execSQL() 方法执行在数据库中创建数据表操作。execSQL() 方法用于对数据库进行无返回数据的 SQL 操作，基本囊括了除查询语句以外的其他 SQL 命令。Score 表中的字段有常见的字符型、数字型和日期型。对于 VARCHAR 字符型字段设置 COLLATE NOCASE，意思是忽略大小写字母的区别。推荐对字段设置默认值，早期很多数据库都默认允许字段值为空值 null，主要是从节省存储空间考虑，而目前存储空间已经不是主

图 5-18 SQLite 数据库运行结果

要问题。取消 null 可简化后续编程的返回值判断。第 24 行建表命令放在 try-catch 中是为了简化判断是否已有 Score 表的逻辑，如果没有 Score 表此行命令就正常运行；如果已经有 Score 表，则捕获异常后继续执行下一条命令。

第 28 行使用 SQL 的 delete 命令删除 Score 表中的数据，保证每次演示开始时都只有后面代码添加的 3 条记录。

第 30～33 行演示用 ContentValues() 加 insert() 方法向 Score 表中添加数据。

第 36～49 行演示用 SQL 的 insert 命令添加两条记录。添加记录的命令是第 39～40 行，多出的代码是为了演示事务处理。execSQL() 一次只能执行一条 SQL 语句，如果执行多条 SQL 语句，就需要执行多次 execSQL() 方法。如果希望所有 execSQL() 方法都执行成功并且提交 SQL 对数据库的修改，如果有一条 SQL 语句执行失败就全部回滚到修改前的状态。为实现上述目标，可以使用事务处理。第 38 行的 beginTransaction() 方法代表事务处理开始。当执行到第 42 行时说明之前的所有 SQL 语句都没有错误，通过 setTransactionSuccessful() 方法通知数据库将之前的修改全部提交。如果有异常产生就转而执行第 45 行进行异常处理。最后执行 48 行调用 endTransaction() 方法结束事务。

第 50 行的 showTable() 是自定义方法，用于显示 Score 表中的记录。

第 53 行开始的 OnClickListener 监听器主要处理单击按钮后数据库的增、删、改操作。需要说明的是，SQLite 数据库的数字字段也是可以保存字符的，只需要将相应 SQL 语句的字段值加上单引号括起来（不建议使用，因为后续的数据处理还需额外判断是否能转换为数字型，增加了数据处理和迁移的难度）。

第 82～109 行是自定义的 showTable() 方法。使用 rawQuery() 方法执行查询命令,返回 Cursor(游标)。查询的后续操作都是基于 Cursor 的方法实现,Cursor 的常用方法如表 5-1 所示。

表 5-1 Cursor 的常用方法

方 法	作 用
getCount()	获取满足条件的记录数
isFirst()	判断是否是第一条记录
isLast()	判断是否是最后一条记录
moveToFirst()	移动到第一条记录
moveToLast()	移动到最后一条记录
move(int offset)	移动到指定记录,正数前移,负数后移
moveToNext()	移动到下一条记录
moveToPrevious()	移动到上一条记录
getColumnName(int columnIndex)	返回 columnIndex 列的字段名
getColumnIndex(String columnName)	返回字段名为 columnName 的列索引
getColumnIndexOrThrow(String columnName)	根据字段名获得列索引,不存在的列抛出异常
getInt(int columnIndex)	获取指定列索引的 int 型值
getString(int columnIndex)	获取指定列索引的 String 型值
getCount()	返回查询结果的记录数
getColumnCount()	返回查询结果的列数

5.10 SQLiteOpenHelper

视频讲解

当 App 升级新版本时可能会对数据库的结构进行调整,如何实现 App 版本升级时都能匹配对应的数据库版本呢? SQLiteOpenHelper 就是用来处理数据库版本升级维护的。

```
【FirstActivity.java】
01    public class FirstActivity extends Activity
02    {
03        @Override
04        public void onCreate(Bundle savedInstanceState)
05        {
06            super.onCreate(savedInstanceState);
07            setContentView(R.layout.main);
08
09            DatabaseHelper databaseHelper = new DatabaseHelper(this);
10            try
11            {
12                //当数据库降级时,此命令会抛出异常
13                SQLiteDatabase myDb = databaseHelper.getWritableDatabase();
14            }
15            catch (Exception e)
16            {
```

```java
17              Log.i("xj", e.toString());
18          }
19          /**
20          //myDb.close();//别忘了在程序退出前关闭数据库连接
21      }
22
23      /**
24       * 用于生成或更新数据库
25       */
26      class DatabaseHelper extends SQLiteOpenHelper
27      {
28          private static final int DATABASE_VERSION = 4;
            //数据库版本。数字变小会引起 databaseHelper.getWritableDatabase()异常
29          TextView textView1 = (TextView) findViewById(R.id.textView1);
30
31          DatabaseHelper(Context context)
32          {
33              super(context, "mydb.db", null, DATABASE_VERSION);
34          }
35
36          //第一次创建数据库时调用
37          @Override
38          public void onCreate(SQLiteDatabase db)
39          {
40              textView1.append("\nonCreate:新建数据库\n" + db.toString());
41          }
42
43          //升级时调用
44          @Override
45          public void onUpgrade(SQLiteDatabase db, int oldVersion, int newVersion)
46          {
47              textView1.append("\nonUpgrade:从版本" + oldVersion + "升级到版本 " +
                    newVersion + "\n" + db.toString());
48          }
49
50          //以下方法都要自行添加
51          //最先调用,所有情况都会调用
52          @Override
53          public void onConfigure(SQLiteDatabase db)
54          {
55              textView1.append("\nonConfigure:");
56              super.onConfigure(db);
57          }
58
59          //最后调用,降级时不调用。为什么
60          @Override
61          public void onOpen(SQLiteDatabase db)
62          {
63              textView1.append("\nonOpen:");
```

```
64                super.onOpen(db);
65            }
66
67            //降级时调用
68            @Override
69            public void onDowngrade(SQLiteDatabase db, int oldVersion, int newVersion)
70            {
71                textView1.append("\nonDowngrade:从版本" + oldVersion + "降级到版本 " +
                    newVersion + "\n" + db.toString());
72                super.onDowngrade(db, oldVersion, newVersion);
73            }
74        }
75    }
```

第 9 行将自定义的 DatabaseHelper 类实例化赋给变量 databaseHelper。

第 13 行调用 getWritableDatabase()方法[如果只读就调用 getReadableDatabase()方法]。此方法会根据版本号分别调用 DatabaseHelper 类中的方法。当版本号从大变到小时,执行此命令会抛出异常。

第 26~74 行定义了继承于 SQLiteOpenHelper 的子类 DatabaseHelper,并重写父类的方法。第 28 行定义常量 DATABASE_VERSION,当数据库有变化时,开发人员可修改源码中的此值,DatabaseHelper 会根据升级 App 前后的 DATABASE_VERSION 值变化分别调用不同的方法。SQLiteOpenHelper 版本变化和调用方法顺序关系如表 5-2 所示。

表 5-2　SQLiteOpenHelper 版本变化和调用方法顺序关系

数据库版本情况	调 用 方 法				
	onConfigure()	onCreate()	onUpgrade()	onDowngrade()	onOpen()
第 1 次运行 App	①	②			③
版本无变化	①				②
版本升级	①		②		③
版本降级	①			②	

数据库版本升降级时会显示变动前后的数据库版本号,开发者需考虑跨版本升级问题。无论什么情况都会调用 onConfigure()方法,当降级时不调用 onOpen()方法。很多开发人员在开发 App 时只使用了 onCreate()和 onUpgrade()方法。读者可根据实际情况将代码放在相应的方法中。

5.11　数据库调试

在开发数据库相关应用时少不了访问数据库以获取开发相关信息。早期最常用的方式是使用 adb shell 命令(如果有多台设备就使用 adb -s emulator-5554 shell 命令,具体的设备名称可以使用 adb devices 命令获取)。操作步骤如下:

(1) 输入 adb shell 命令进入 Android 模拟器命令行界面。

(2) 输入 su 命令进入 root 模式(真实的 Android 设备要进入 root 模式以后才行)。

视频讲解

(3) 输入 cd /data/data/com.xiaj/databases/ 命令进入 mydb.db 所在目录。

(4) 输入 sqlite3 mydb.db 命令进入 mydb.db 数据库。此时提示符变为 sqlite>。

(5) 输入 .table 命令可以查看数据表。注意，table 前有一个点。

(6) 输入"select * from Score;"语句可查询表记录。

输入命令级显示结果如下所示：

```
D:\> adb -s emulator-5554 shell
generic_x86_64:/ $ su
generic_x86_64:/ # cd /data/data/com.xiaj/databases/
generic_x86_64:/data/data/com.xiaj/databases # ls
mydb.db mydb.db-journal
generic_x86_64:/data/data/com.xiaj/databases # sqlite3 mydb.db
SQLite version 3.22.0 2018-12-19 01:30:22
Enter ".help" for usage hints.
sqlite> .table
Score                android_metadata
sqlite> .schema
CREATE TABLE android_metadata (locale TEXT);
CREATE TABLE Score (Id INTEGER PRIMARY KEY, Name VARCHAR (30) NOT NULL DEFAULT '' COLLATE NOCASE,
Score FLOAT NOT NULL DEFAULT 0, InDate DATETIME NOT NULL DEFAULT(DATETIME('now', 'localtime') ) );
sqlite> sqlite> select * from Score;
1|刘备|95.0|2021-02-19 09:11:48
2|关羽|80.0|2021-02-19 09:11:48
3|张飞|90.0|2021-02-19 09:11:48
sqlite>
```

可以输入 inert、delete、update 命令对数据表中的记录执行增、删、改操作。但此种方法有两个缺点：

(1) 进入 /data/data 目录需要 root 权限。真实的 Android 设备需要刷机才能获取 root 权限，但也意味着失去厂商保修资格。

(2) 当输入的 SQL 命令中包含中文时会显示为乱码（早期版本的 Windows 10 还需先运行 chcp 65001 命令才能在查询结果中显示中文，最新版本的 Windows 10 无须再输此命令）。

简单的替代方法是开发人员在 App 中单独开发一个 Activity，输入预置的 SQL 语句来查询数据。但如果面对的是复杂多样的查询或数据表结构改变操作，此方法有点力不从心。此时各式各样的插件应运而生，有的插件在 Android 应用中提供 Web 服务，开发人员可以通过浏览器访问数据库，此类插件的实时交互性有待改善。有的插件在 Android 中提供中转服务，再由客户端提供图形化实时交互操作界面，典型的代表就是 SQLiteStudio。以下以 SQLiteStudio 方式讲解实时管理 Android 设备上的 SQLite 数据库。

打开 SQLiteStudio 应用程序，选择"工具"→"打开配置对话框"菜单，如图 5-19 所示。

在弹出的配置对话框中选择"插件"选项，选中 Android SQLite 复选框，如图 5-20 所示，单击 OK 按钮。

此时"工具"菜单中多出一项 Get Android connector JAR file，如图 5-21 所示。

单击此菜单项下载 SQLiteStudioRemote.jar 文件。

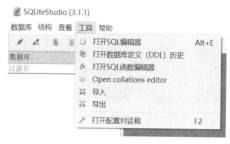

图 5-19　选择"工具"→"打开配置对话框"菜单

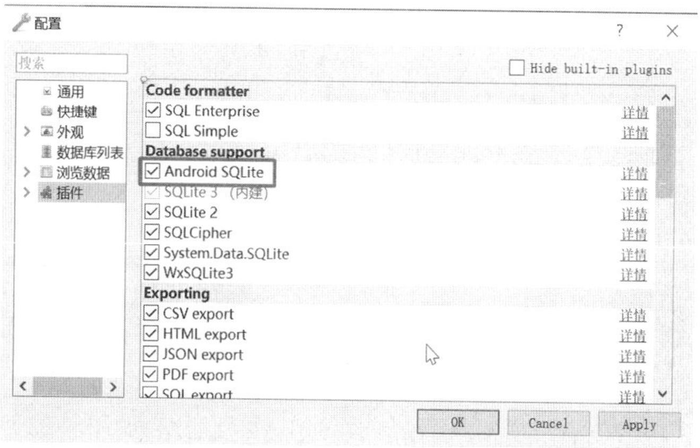

图 5-20　选中 Android SQLite 复选框

图 5-21　下载 .jar 文件

将下载的 SQLiteStudioRemote.jar 文件放到 Android 项目的 libs 目录下（如果没有就新建 libs 目录）。在 app 目录下的 build.gradle 文件中添加如下配置（具体路径视文件所在路径而定）：

```
implementation files('src/main/libs/SQLiteStudioRemote.jar')
```

在 Activity 的 onCreate() 方法中添加以下代码：

```
SQLiteStudioService.instance().start(this);
```

此时运行 Android 项目，SQLiteStudio 的服务就同步启动。

最后配置 SQLiteStudio 连接 App 的数据库。在 SQLiteStudio 软件中选择"添加数据库"菜单,在"数据库"对话框中选择数据类型为 Android SQLite。单击 Android database URL 图标,在弹出的对话框中选中 USB cable-port forwarding 单选按钮,如图 5-22 所示,单击 OK 按钮。

图 5-22　配置数据库连接

连接数据库成功后就可以实时查看、修改 SQLite 数据库,如图 5-23 所示。

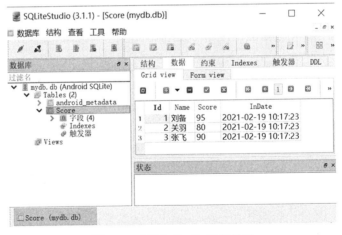

图 5-23　SQLiteStudio 运行界面

SQLiteStudio 软件的数据库管理功能非常完善,完全能满足开发要求,连接真实的 Android 设备也没有问题。

Android Studio 从 4.1 版本开始引入 Database Inspector 功能,可以连接 API level 26 以上版本的 SQLite 数据库连接。选择 View→Tool Windows→Database Inspector 菜单,在下方出现 Database Inspector 选项卡,运行 App 程序,出现如图 5-24 所示的界面。界面左边显示数据库及所包含的表。单击 ![btn] 按钮,界面右侧出现查询选项卡,在文本框中输入 SQL 语句,单击 Run 按钮执行相应的 SQL 命令。数据的变更可以实时显现(单击 Run 按钮可实时查询表记录。如果选中图 5-24 的 Live updates 复选框，App 中对表数据的修改会自动更新显示）。此方案是目前连接数据库并实现实时查询操作的最简单方案。但功能上与 SQLiteStudio 方案相比还有较大的差距（如在图形化界面中修改表结构等）。

图 5-24　Database Inspector

【注】　目前的 Android Studio 版本使用 Database Inspector 功能时，Activity 中的 onCreate()方法中最好保持数据库连接打开。如果 onCreate()方法中对数据库执行了关闭操作,需要后续执行打开数据库操作才能显示出数据库和表。从实际运行看,程序启动后过几秒才连接到数据库,在 onCreate()方法中关闭数据库连接会导致 Database Inspector 无法找到数据库。

5.12　SharedPreferences

视频讲解

有时需要判断用户是第几次启动 App,如果是第一次启动,就要先显示广告或软件功能介绍界面,或者 App 试用版启动试用次数限制功能。解决上述设计需求的一种方案是开发人员自己设定一个变量记录启动次数,然后保存到 SD 卡,每次启动 App 时都判断此变量的值。此方案需要 SD 卡的读写权限,并需要添加大量代码完成 SD 卡的读写,一种更简便的方案是使用 SharedPreferences。SharedPreferences 是一个轻量级的存储类,提供简单数据类型和包装类的数据存取接口。

```
【FirstActivity.java】
01    public class FirstActivity extends Activity
02    {
03        @Override
04        protected void onCreate(Bundle savedInstanceState)
05        {
06            super.onCreate(savedInstanceState);
07            setContentView(R.layout.main);
08
09            TextView textView1 = (TextView) findViewById(R.id.textView1);
```

```
10      SharedPreferences sharedPreferences = getSharedPreferences("MyApp", Context.MODE_
        PRIVATE);
11
12      boolean isFirst = sharedPreferences.getBoolean("IsFirst", true);
13      SharedPreferences.Editor editor = sharedPreferences.edit();
14      if (!isFirst)
15      {
16          editor.putBoolean("IsFirst", fals);
17          editor.putString("Name", "灰太狼");
18          editor.putInt("LoginCount", 1);
19          editor.commit();            //保存更新
20
21          textView1.append("这是" + sharedPreferences.getString("Name", "") +
            "第一次光临");
22      }
23      else
24      {
25          int loginCount = sharedPreferences.getInt("LoginCount", 1) +1;
26          editor.putInt("LoginCount", loginCount);
27          editor.commit();            //保存更新,否则还是2
28          textView1.append(sharedPreferences.getString("Name", "") + "又回来
            了!这是第" + loginCount + "次了.");
29      }
30  }
31 }
```

第 10 行生成一个 sharedPreferences 对象实例,其存储数据区域取名 MyApp,后续保存的键值对都放在此区域。

第 12 行调用 sharedPreferences 中的 getBoolean()方法来获取键值对中名为 IsFirst 的对应值,如果查不到 IsFirst 的对应值就取第二个参数 true 作为默认值。

第 13 行调用 edit()方法,返回的对象赋予变量 editor,便于后续对 editor 的连续调用完成保存键值对操作。

第 14 行判断变量 isFirst,如果条件成立代表是第一次启动 App,程序从第 16 行开始执行,使用 putBoolean()、putString()、putInt()方法将布尔型、字符串和整型值传递给相应的 key。如果条件不成立则从第 25 行开始执行,将登录次数值执行加 1 操作后提交 editor 的修改。

第 19 行是将变更后的键值对保存到当前应用程序所属储存空间,所以即使关闭 App 或关机也不影响键值对的内容,只有卸载 App 时应用程序所属储存空间才会被删除。

5.13 精度问题

以下是运行 1.2-0.1 的 Log.i()方法及输出结果:

```
Log.i("xj","1.2 - 0.1 =" + (1.2 - 0.1));
I: 1.2 - 0.1 = 1.0999999999999999
```

输出结果不是我们认为的 1.1,究其原因在于十进制数 1.2 在计算前需要转为二进制数,其中十进制的小数部分 0.2 转换为二进制小数会变为无限循环小数 $0.00\dot{1}\dot{1}$,而 Java 中双精度数是用 64b 表示的,意味着转换为二进制后会有精度丢失。本案例演示了如下两种精度处理方案。

（1）使用 DecimalFormat 设置保留小数点后位数。
（2）使用 BigDecimal 进行计算。

【FirstActivity.java】
```
01  public class FirstActivity extends Activity
02  {
03      @Override
04      public void onCreate(Bundle savedInstanceState)
05      {
06          super.onCreate(savedInstanceState);
07          setContentView(R.layout.main);
08
09          TextView textView1 = (TextView) findViewById(R.id.textView1);
10          Button button1 = (Button) findViewById(R.id.button1);
11          Button button2 = (Button) findViewById(R.id.button2);
12          Button button3 = (Button) findViewById(R.id.button3);
13
14          textView1.setText("各数据类型范围: ");
15          textView1.append("\nByte.MAX_VALUE:" + Byte.MAX_VALUE);
16          textView1.append("\nByte.MIN_VALUE:" + Byte.MIN_VALUE);
17          textView1.append("\nShort.MAX_VALUE:" + Short.MAX_VALUE);
18          textView1.append("\nShort.MIN_VALUE:" + Short.MIN_VALUE);
19          textView1.append("\nInteger.MAX_VALUE:" + Integer.MAX_VALUE);
20          textView1.append("\nInteger.MIN_VALUE:" + Integer.MIN_VALUE);
21          textView1.append("\nLong.MAX_VALUE:" + Long.MAX_VALUE);
22          textView1.append("\nLong.MIN_VALUE:" + Long.MIN_VALUE);
23
24          textView1.append("\nFloat.MAX_VALUE:" + Float.MAX_VALUE);
25          textView1.append("\nFloat.MIN_VALUE:" + Float.MIN_VALUE);
26          textView1.append("\nDouble.MAX_VALUE:" + Double.MAX_VALUE);
27          textView1.append("\nDouble.MIN_VALUE:" + Double.MIN_VALUE);
28
29          button1.setOnClickListener(new OnClickListener()
30          {
31              @Override
32              public void onClick(View v)
33              {
34                  textView1.setText("1.2 - 0.1 = " + (1.2 - 0.1));
35                  textView1.append("\n\n1/3 + 1/3 + 1/3 = " + (1 / 3 + 1 / 3 + 1 / 3));
36              }
37          });
38
39          button2.setOnClickListener(new OnClickListener()
40          {
41              @Override
```

```java
42              public void onClick(View v)
43              {
44
45                  textView1.setText("DecimalFormat 方法指定的 3 位精度:1.2 - 0.1 = " +
                        decimalFormatPrecision(1.2 - 0.1));
46                  textView1.append("\n\nDecimalFormat 超出 3 位精度则四舍五入: 1.1 -
                        0.01 = " + decimalFormatPrecision(1.1 - 0.01));
47
48                  textView1.append("\n\nBigDecimal: 1.2 - 0.1 = " + sub("1.2", "0.1"));
49                  textView1.append("\n\nBigDecimal: 1.1 - 0.01 = " + sub("1.1", "0.01"));
50                  textView1.append("\n1.0/3.0 + 1.0/3.0 + 1.0/3.0 = " + (1.0 / 3.0 +
                        1.0 / 3.0 + 1.0 / 3.0));
51                  textView1.append("\n\nBigDecimal: 1/3 + 1/3 + 1/3 = " + (div(1, 3) +
                        div(1, 3) + div(1, 3)));
52              }
53          });
54
55          button3.setOnClickListener(new OnClickListener()
56          {
57              @Override
58              public void onClick(View v)
59              {
60                  textView1.setText("new BigDecimal(2.3) = " + new BigDecimal(2.3));
61                  textView1.append("\n\nnew BigDecimal(\"2.3\") = " + new BigDecimal
                        ("2.3"));
62                  textView1.append("\n\nBigDecimal.valueOf(2.3) = " + BigDecimal.
                        valueOf(2.3));
63              }
64          });
65      }
66      double decimalFormatPrecision(double n)
67      {
68          //DecimalFormat 需要 API 24 以上才支持
69          DecimalFormat decimalFormat = new DecimalFormat("0.###");
70          return Double.parseDouble(decimalFormat.format(n));
71      }
72
73      BigDecimal sub(String num1, String num2)
74      {
75          BigDecimal bd1 = new BigDecimal(num1);
76          BigDecimal bd2 = new BigDecimal(num2);
77          return bd1.subtract(bd2);
78      }
79
80      double div(double num1, double num2)
81      {
82          int scale = 16; //精度 16(含)位以上即可
83          BigDecimal bd1 = new BigDecimal(Double.toString(num1));
84          BigDecimal bd2 = new BigDecimal(Double.toString(num2));
85          //return bd1.divide(bd2, scale, BigDecimal.ROUND_HALF_UP).doubleValue();
86          return bd1.divide(bd2, scale, RoundingMode.HALF_UP).doubleValue();
87      }
88  }
```

第 15~27 行显示不同数据类型的最小值和最大值。运行结果如图 5-25 所示。

第 29~37 行 button1 的单击监听器演示双精度小数减法和整型除法结果再相加计算。运行结果如图 5-26 所示。第 34 行的结果误差来源于进制转换代码的精度丢失。第 35 行的输出结果是 0,原因是 int 型的 1 除以 3 结果为取整后的 0,三个 0 相加还是 0。

第 39~53 行演示对精度的常用处理方式,如指定保留位数的四舍五入和使用 BigDecimal 类的精度计算处理。

第 45 和 46 行调用第 66~71 行的自定义方法 decimalFormatPrecision()来处理精度,其中使用 DecimalFormat 来设定数值的精度保留位数。此方案对不同小数位数的计算显得灵活性不够,会因为保留小数位数较短导致计算结果精度丢失。

第 48~49 行调用第 73~78 行的自定义方法 sub()实现 BigDecimal 的减法操作。为保证计算结果的精度,原则上将 BigDecimal 转回相应的数据类型的时机尽可能后延。

第 50 行使用 double 型的数值进行除法和加法运算。

第 60 行将 double 型的 2.3 作为 BigDecimal 的构造方法实参,其输出结果有精度丢失。

第 61 行将 String 型的 2.3 作为 BigDecimal 的构造方法实参,其输出结果没有精度丢失。

第 62 行使用 BigDecimal 的 valueOf()方法也能保证精度。

程序运行结果如图 5-25~图 5-28 所示。

图 5-25 数值范围

图 5-26 未处理精度计算结果

第 85 行的 BigDecimal.ROUND_HALF_UP 已被 Java 弃用,改为第 86 行的 RoundingMode.HALF_UP。RoundingMode 舍入含义如表 5-3 所示。RoundingMode 舍入保留个位数样例如表 5-4 所示。

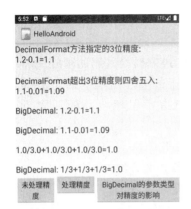

图 5-27　处理精度问题

图 5-28　参数类型对精度的影响

表 5-3　RoundingMode 舍入含义

舍入关键字	舍入方式
UP	向远离 0 的方向舍入
DOWN	向 0 方向舍入
CEILING	向正无穷方向舍
FLOOR	向负无穷方向舍入
HALF_UP	四舍五入
HALF_DOWN	五舍六入
HALF_EVEN	四舍六入，五留双
UNNECESSARY	计算结果是精确的，不需要舍入模式，会抛出 ArithmeticException 异常

表 5-4　RoundingMode 舍入保留个位数样例

数字	UP	DOWN	CEILING	FLOOR	HALF_UP	HALF_DOWN	HALF_EVEN	UNNECESSARY
5.5	6	5	6	5	6	5	6	抛出 ArithmeticException 异常
2.5	3	2	3	2	3	2	2	抛出 ArithmeticException 异常
1.6	2	1	2	1	2	2	2	抛出 ArithmeticException 异常
1.1	2	1	2	1	1	1	1	抛出 ArithmeticException 异常
1	1	1	1	1	1	1	1	1
−1	−1	−1	−1	−1	−1	−1	−1	−1
−1.1	−2	−1	−1	−2	−1	−1	−1	抛出 ArithmeticException 异常
−1.6	−2	−1	−1	−2	−2	−2	−2	抛出 ArithmeticException 异常
−2.5	−3	−2	−2	−3	−3	−2	−2	抛出 ArithmeticException 异常
−5.5	−6	−5	−5	−6	−6	−5	−6	抛出 ArithmeticException 异常

5.14 横 竖 屏

有时设备上的屏幕自动旋转功能可能被用户关闭,而 App 的 Activity 可能需要横屏显示。本案例演示通过代码设置横屏或竖屏显示。

【FirstActivity.java】
```java
01  public class FirstActivity extends Activity
02  {
03      @Override
04      public void onCreate(Bundle savedInstanceState)
05      {
06          super.onCreate(savedInstanceState);
07          setContentView(R.layout.main);
08          Button button1 = (Button) this.findViewById(R.id.Button1);
09          Button button2 = (Button) this.findViewById(R.id.Button2);
10
11          Point point = new Point();
12          getWindowManager().getDefaultDisplay().getSize(point);
13          Log.i("xj","\n设备分辨率为: " + point.toString());
14
15          button1.setOnClickListener(new OnClickListener()
16          {
17              public void onClick(View arg0)
18              {
19                  setRequestedOrientation(ActivityInfo.SCREEN_ORIENTATION_LANDSCAPE);
                    //横屏
20              }
21          });
22
23          button2.setOnClickListener(new OnClickListener()
24          {
25              public void onClick(View arg0)
26              {
27                  setRequestedOrientation(ActivityInfo.SCREEN_ORIENTATION_PORTRAIT);
                    //竖屏
28              }
29          });
30      }
31
32      @Override
33      public void onConfigurationChanged(Configuration config)
34      {
35          super.onConfigurationChanged(config);
36          if (config.orientation == Configuration.ORIENTATION_LANDSCAPE)
37          {
38              Log.i("xj","\n现在是横屏");
39          } else if (config.orientation == Configuration.ORIENTATION_PORTRAIT)
```

```
40          {
41              Log.i("xj","\n现在是竖屏");
42          }
43          Log.i("xj", "onConfigurationChanged:" + config.toString());
44      }
45  }
```

第 11～13 行是获取屏幕分辨率并在 Logcat 中输出。

第 19 行将 App 设为横屏显示，第 27 行将 App 设为竖屏显示。

第 33～44 行重写 onConfigurationChanged() 方法，其中对象变量 config 的 orientation 属性代表当前设备是处于横屏还是竖屏状态。

当运行程序时，单击两个按钮会将屏幕设为相应的横屏或竖屏，同时会调用当前 Activity 的 onCreate() 方法。Logcat 中的输入如下：

```
I: 设备分辨率为: Point(1080, 1794)
I: 设备分辨率为: Point(1794, 1080)
I: 设备分辨率为: Point(1080, 1794)
```

可以看出每次改变横竖屏都会执行 onCreate() 方法并执行第 13 行的 Log.i() 方法输出分辨率。注意观察横竖屏时分辨率的变化。

在 AndroidManifest.xml 文件的 Activity 标签中添加以下属性：

android:configChanges = "orientation|screenSize"

再次运行程序并单击按钮，输出结果如下：

```
I: 设备分辨率为: Point(1080, 1794)
I: 现在是横屏
I: onConfigurationChanged:{1.0 310mcc260mnc [zh_CN #Hans,en_US] ldltr sw411dp w683dp h387dp
420dpi nrml land finger qwerty/v/v - nav/h winConfig = { mBounds = Rect(0, 0 - 1920, 1080)
mAppBounds = Rect(0, 0 - 1794, 1080) mWindowingMode = fullscreen mDisplayWindowingMode =
fullscreen mActivityType = standard mAlwaysOnTop = undefined mRotation = ROTATION_90} s.3}
```

当横竖屏变化时不再调用 onCreate() 方法，而是调用第 33～44 行的 onConfigurationChanged() 方法。此方式可防止重复调用 onCreate() 方法导致控件数据被重新初始化。

5.15 获取 App 信息

在判断 App 是否需要升级等应用场景，首先需要获取当前 App 的版本信息。本案例演示获取 App 的信息。

【FirstActivity.java】
```
01  public class FirstActivity extends Activity
02  {
03      @Override
```

```
04      public void onCreate(Bundle savedInstanceState)
05      {
06          super.onCreate(savedInstanceState);
07          setContentView(R.layout.main);
08
09          TextView textView1 = (TextView) findViewById(R.id.textView1);
10          PackageManager packageManager1 = getPackageManager();
11
12
13          try
14          {
15              PackageInfo packageInfo = packageManager1.getPackageInfo
                    (getPackageName(), 0);
16              textView1.append("\nPackageName:" + packageInfo.packageName);
17              textView1.append("\nVersionCode:" + packageInfo.versionCode);
18              textView1.append("\nVersionName:" + packageInfo.versionName);
19              textView1.append("\nVersionName:" + packageInfo.applicationInfo);
20              textView1.append("\nVersionName:" + packageInfo.firstInstallTime);
21              textView1.append("\nVersionName:" + packageInfo.lastUpdateTime);
22              textView1.append("\nVersionName:" + packageInfo.toString());
23          } catch (NameNotFoundException e)
24          {
25              e.printStackTrace();
26          }
27      }
28  }
```

第 10 行通过 getPackageManager()方法获取当前 App 的 PackageManager 对象实例。

第 16~22 行获取当前 App 的包名、版本号、版本名称、第一次安装时间、上次升级时间等信息。获取 App 信息结果如图 5-29 所示。

图 5-29 获取 App 信息结果

附录 A 综合实验

1. 使用所学知识编写一个计算器 App。
2. 竖屏时能实现基本的加减乘除运算、回退和清空输入。横屏时变为科学计算器，实现函数计算、进制换算等功能。
3. 数字按键依照手机拨号键盘顺序排列。
4. 选择相应菜单可对计算器相关特性进行配置，如选择进制转换、是否全屏显示、选择横竖屏、显示区文字大小、颜色等。
5. 加号和等号外观尺寸应比其他正常按键大。
6. 输入计算公式，按等号按钮后按优先级计算表达式并输出结果。
7. 结果显示区应支持长按弹出复制、粘贴功能。
8. 使用计算器过程中不允许弹出软键盘。
9. 可以通过进度条实时调整计算结果保留的小数点后位数，或者通过音量键完成同样的效果。
10. 运行结果应精准，对错误的计算公式应进行容错处理，至少要考虑如下计算情况。

$1/3+1/3+1/3$

$1.1-1.001$

$1/0$、$1/(1-1)$ 和 $1/\sin(0)$

$3/2$

$\sqrt[2]{-1}$

参 考 文 献

[1] 王翠屏. Android Studio 应用开发实战详解[M]. 北京：人民邮电出版社，2017.
[2] 杨明羽. Android 语法范例参考大全[M]. 北京：电子工业出版社，2012.
[3] 兰红，李淑芝. Android Studio 移动应用开发从入门到实践[M]. 北京：清华大学出版社，2018.

图书资源支持

感谢您一直以来对清华版图书的支持和爱护。为了配合本书的使用,本书提供配套的资源,有需求的读者请扫描下方的"书圈"微信公众号二维码,在图书专区下载,也可以拨打电话或发送电子邮件咨询。

如果您在使用本书的过程中遇到了什么问题,或者有相关图书出版计划,也请您发邮件告诉我们,以便我们更好地为您服务。

我们的联系方式:

地　　址:北京市海淀区双清路学研大厦A座714

邮　　编:100084

电　　话:010-83470236　010-83470237

客服邮箱:2301891038@qq.com

QQ:2301891038(请写明您的单位和姓名)

资源下载: 关注公众号"书圈"下载配套资源。

书 圈

获取最新书目

观看课程直播